Think Control, Not Risk!

Mastering Health and Safety under the Model WH&S Act

A Revolution in Safety Thinking for Directors, Managers and Professionals in Small, Medium and Large Organisations

Marcus Punch

1st Edition

1st Edition

ISBN 978-0-9807660-3-5 (paperback)
RRP $AUD 29.95 (including GST), plus postage and handling.

Published by:
Marcus Punch Pty. Ltd., Tenambit NSW, Australia
Email: marcus@marcuspunch.com
Web: www.marcuspunch.com
Phone: +61 (0)432 168849

The first step to a successful revolution is
destroy all competing revolutionaries.

— lyrics from "Jehovah Made this Whole Joint for You", by *The New Radicals – Maybe You Have Been Brainwashed Too* (1998).

Table of Contents

List of Tables and Figures

Chapter 1 - Myths, Snake Oil & Alchemy

Since the Model Work Health and Safety Act and Regulations were introduced in 2011 much has been written about their requirements and implications for those involved in managing safety at work and in workplaces.

Whilst a good deal of reliable support material has been provided by the regulatory authorities, a good deal of conflicting information has emanated from other, sometimes questionable, sources.

The conflicting information is usually related to the meaning of certain aspects of the Model Act and Regulations, their intended application, who is liable to prosecution, or the process by which the requirements must be satisfied (ie. what I have coined "myths, snake oil and alchemy"). This book provides a straight-forward, stand-alone "big picture" view for those running small, medium and large organisations, managers and professionals and those with specific duties or responsibilities related to ensuring safety under the Model Act and Regulations.

There have also been perceptions in various quarters about the suitability of risk management standards such as AS/NZS ISO31000 for safety-related risk management – particularly where 'level of risk' is used as a basis for making decisions about accepting / tolerating a risk. The Model Act and Regulations take a different approach - that a decision to accept / tolerate a risk should only be made when all "reasonably practicable" risk controls are in place.

Unfortunately, very little has been said by the regulatory authorities to clarify this matter. A key aim of this book is to clarify, rationalise and enhance risk management standards such as AS/NZS ISO31000 in the context of the Model Act and Regulations.

The mantra of this book is: **Think control, not (level of) risk'!**

It is not intended as a comprehensive "how to" for safety management under the Model Act. Much has already been published in this regard by the regulatory authorities. Indeed, in safety management, the details of the "how to" often vary from application to application – depending on the nature of the work or workplace being managed and other factors – be it a railway, airline, mine, retail business, hospital or a volunteer organisation such as a scout troop or a community garden. Nevertheless, the book provides references to reliable, authoritative resources which provide detail on the various aspects of the "how to".

Instead, this book aims to:

- Explain the essential concepts, principles and requirements for ensuring health and safety that are contained in the Model Act.
- Relate them to the established frameworks for health and safety management and risk management.
- Expose authoritative sources (and unauthoritative ones) for interpretation and learned opinions on the subject matter.
- Provide a critique on a number of contemporary health and safety management approaches through the prism of the Model Act's principles, concepts and requirements.
- Propose an enhanced framework for health and safety-related risk management,
- and in doing so, provide the means by which readers can critically evaluate information that they might receive on the subject.

Chapter 2 - The Model Act and Regulations

In 2011 work health and safety legislation was reformed in Australia. Previously, each state of Australia had taken responsibility for enacting its own work health and safety laws via the provisions of the Australian Constitution. This contributed to differences in the safety performance achieved across the various jurisdictions as well as inefficiencies and loss of productivity in many organisations which operate in multiple jurisdictions across Australia. The goal of the National Review of Occupational Health and Safety Laws was to develop a nationally consistent Act and Regulation framework that all Australian jurisdictions would accept and adopt.

This harmonisation of Australian work health and safety laws has been part of the Council of Australian Governments (COAG) National Reform Agenda. The *Model Workplace Health and Safety Bill 2011* (the Model Act) was the culmination of a long review process that was commissioned in 2008. The Model Act was accompanied by Model Regulations.

The Model Act largely follows the architecture of the post-Robens[1] style workplace health and safety statutes that have evolved in the Commonwealth and state jurisdictions in Australia since the 1980s.

It was intended that all Australian states would enact 'mirror' legislation, adopting the provisions of the Model Act by 1st January 2012. At the time of publication of this book, all states with the exception of Western Australia and Victoria, had adopted the Model Act and Model Regulations. However, many of the key concepts and principles of the Model Act also apply in the jurisdictions of Western Australia and Victoria

---

[1] Robens, Lord A, 1972, *Report of the Committee on Safety and Health at Work, 1970-1972*, HMSO, London, Cmnd 5034.

where post-Robens style workplace health and safety legislation is also in place.

Following a multiple fatality mine disaster at the Pike River coal mine in 2012, New Zealand has undertaken a widespread revision of its own legislation. Many of the concepts and features of the Australian Model Act and Regulations have been adopted in the *New Zealand Health and Safety Reform Bill 2014.*

The following is a summary of the key concepts, principles and requirements of the Model Act (and Regulations).

Person Conducting a Business or Undertaking[2]

The Model Act defines a number of duty holders, of which the most significant are:

- A "person conducting a business or undertaking" (also known as a PCBU).
- An "officer of a person conducting a business or undertaking" (covered in the next section).

Importantly, PCBU has been used instead of the term 'employer' and 'worker' has been used instead of 'employee'. This change in terminology has significantly expanded the extent of the duties and obligations under the Model Act.

---

[2] See the following references for further details about PCBUs - Safe Work Australia, 2011, *Interpretative Guideline "The Meaning of Person Conducting a Business or Undertaking"*, Workcover NSW, 2011, *Fact Sheet "PCBUs, Workers and Officers"* and Johnstone, R. and Tooma, M., 2012, *"Work Health and Safety Regulation in Australia – The Model Act"*, The Federation Press.

A PCBU is defined in the Model Act (Section 5)[3] as a 'person' conducting a 'business' or 'undertaking', alone or with others, whether or not for profit or gain.

The Model Act does not define the terms 'business' or 'undertaking'. However, in general terms, a 'business' is an enterprise usually conducted with a view to making a profit and has a degree of organisation, system and continuity. An 'undertaking' may have elements of organisation, systems, and possibly continuity, but is usually not profit-making or commercial in nature.

A 'person', in law, includes a natural person / individual (ie. a human), a body corporate (eg. a company or trust), an unincorporated body or association, a partnership and the Crown.

An individual is only a PCBU where they are conducting a business in their own right (eg. as a sole trader or self-employed person). Individuals who are in a partnership that is conducting a business or undertaking will individually and collectively be a PCBU. The Model Act makes it clear that an individual is not a PCBU if they are involved in the business or undertaking only as a worker or officer.

The Model Act and Regulations also identify specific organisations that are not considered to be PCBUs for the purposes of the Model Act. These organisations include:

- Volunteer associations.
- Elected members of local authorities, who are acting in that capacity.

---

[3] Each state has enacted its own work health and safety legislation based on the Model Act. As such, the section numbering in the Model Act may not exactly reflect the section numbering in each state's legislation.

- Strata title bodies corporate that are responsible for common areas of residential properties, unless the strata title body corporate engages one or more workers as employees.
- Individual householders who engage persons to carry out adhoc home maintenance and repairs or other domestic work. However, an individual householder *may* be a PCBU if they *engage* a worker (eg. employing a nanny to care for children in the householder's home). Furthermore, a householder may also be a PCBU if 'work' is carried out by or for them that is not purely domestic (e.g. work carried out for a business or undertaking that is operated from the householder's home).

Officer of a PCBU[4]

In the Model Act, an officer of a PCBU has the meaning given in Section 9 of the *Corporations Act 2001 (Commonwealth)*, with the exception that each partner within a partnership is not an officer of a PCBU. Each partner within a partnership is a PCBU in their own right.

An officer is a person who makes, or participates in making, decisions that affect the whole, or a substantial part, of the organisation's activities (e.g. a director, company secretary, general manager). A worker, including a member of a profession (e.g. an engineer, lawyer, accountant, teacher, doctor, etc…) or a manager in an organisation is unlikely to be an officer unless they make or participate in making, decisions

---

[4] See the following references for further details about officers of PCBUs - Safe Work Australia, 2011, *Interpretative Guideline "The Health and Safety Duty of an Officer Under Section 27"*, Workcover NSW, 2011, *Fact Sheet "PCBUs, Workers and Officers"* and Johnstone, R. and Tooma, M., 2012, "Work Health and Safety Regulation in Australia – The Model Act", The Federation Press.

that affect the whole, or a substantial part, of the business or undertaking.

An officer of the Crown or a public authority can also be an officer of a PCBU if they are a person who makes, or participates in making, decisions that affect the whole, or a substantial part, of the business or undertaking of the Crown or public authority.

However, an elected member of a municipal council or a minister of a state, territory or the Commonwealth, when acting in that capacity, is not an officer of a PCBU.

Therefore, only a small proportion of the people working within an organisation would be classed as officers of a PCBU (eg. the directors, the company secretary, the general manager, the design/engineering manager of a PCBU involved in design, the construction manager for a PCBU involved in construction etc…).

The Primary Duty of a PCBU[5]

The Model Act (Section 19) sets out the primary duty of a PCBU, which is to ensure, "so far as is reasonably practicable", the health and safety of workers and others. The term "so far as is reasonably practicable" is often abbreviated to "SFARP" or "SFAIRP". Failure to comply with this duty is a criminal offence, and a PCBU can be prosecuted for not doing so.

[5] See the following references for further details about PCBUs - Safe Work Australia, 2011, *Interpretative Guideline "The Meaning of Person Conducting a Business or Undertaking"*, Workcover NSW, 2011, *Fact Sheet "PCBUs, Workers and Officers"* and Johnstone, R. and Tooma, M., 2012, "Work Health and Safety Regulation in Australia – The Model Act", The Federation Press.

'Health' has been defined broadly in the Model Act to mean physical and psychological health, which includes psycho-social health risks such as stress, fatigue and bullying.

PCBUs have a duty to protect 'workers' and 'other persons' from harm, but this has been qualified by the concept of "so far as is reasonably practicable" (which is explained in the next section).

The 'worker' is a natural person / individual who carries out 'work' in any capacity for a PCBU.

'Work' is not defined in the Model Act but is given its ordinary meaning. The following criteria may assist in determining if an activity is 'work' for the purposes of the Model Act:

- The activity involves physical or mental effort by a person or the application of particular skills for the benefit of another person or for themselves (if self-employed), whether or not for profit or payment.
- Activities for which the person or other people will ordinarily be paid by someone is likely to be considered to be work.
- Activities that are part of an ongoing process or project may all be work if some of the activities are for remuneration.
- An activity may be more likely to be work where control is exercised over the person carrying out the activity by another person.
- Formal, structured or complex arrangements may be more likely to be considered to be work than adhoc or unorganised activities.

A 'workplace' is a place where work is carried out for a business or undertaking and includes any place where a worker goes, or is likely to be, while at work.

Importantly, the PCBU's duty encompasses a duty to 'other persons'. A PCBU must ensure, so far as is reasonably practicable, that the health and safety of other persons is not put at risk from work carried out as part of the conduct of the business or undertaking.

'Other persons' may include persons who are located at the workplace but do not perform work (e.g. visitors), or persons located near the workplace, but not involved in working there (e.g. neighbours) who may be exposed to the hazards / risks associated with work carried out at the workplace (e.g. emission of dust or toxic substances, fire or explosions etc...).

The primary duty of a PCBU is shown diagrammatically in Figure 2.1 (within the dotted line). This duty is distinct from the duty of an officer of a PCBU (which is described in a later section).

Figure 2.1: Primary Duty of a PCBU

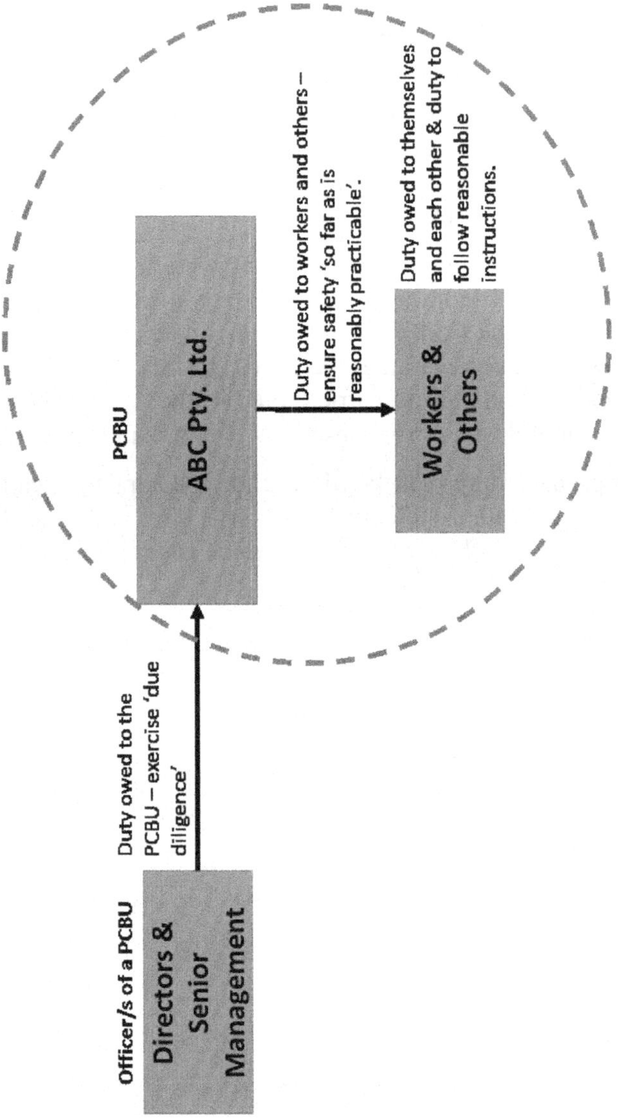

The Model Act also places duties on PCBUs who carry out the following activities associated with work or workplaces:

- Management or control of a workplace (Section 20).
- Management or control of fixtures, fittings or plant at a workplace (Section 21).
- Design, manufacture, import or supply of plant, structures or substances that are, or could be used at a workplace (Sections 22-25).
- Install, construct or commission plant or structures that are, or could be used at a workplace (Section 26).

For instance, a PCBU who designs plant, substances or structures must ensure, so far as is reasonably practicable, that the plant, substance or structure is designed to be without risks to the health or safety of workers and others.

In support of the PCBU's primary duties, the Model Act and Model Regulations (Chapter 3) set out further duties for PCBUs in relation to the identification and management of risks, which are to

- Identify reasonably foreseeable hazards (Model Regulation Section 34).
- Manage risks to health and safety (Model Act Section 17 and Model Regulations Section 35).
- Use the hierarchy of control measures (Model Regulation Section 36).

These duties have also been qualified by the concept of "so far as is reasonably practicable".

The Meaning of Reasonably Practicable[6]

The concept of "reasonably practicable" arose from a UK High Court case, known as *Edwards v National Coal Board*[7]. The case concerned the collapse of an underground roadway in a coal mine which killed a Mr. Edwards.

The National Coal Board argued that it was too expensive to shore up every roadway in all of its mines. The High Court decided that not all of the roadways needed shoring up, just the ones that actually required it. This case established that the extent of the effort required to control a risk would be moderated by the cost, time and trouble involved.

Justice Asquith stated in his judgment:

"Reasonably practicable is a narrower term than 'physically possible' and implies that a computation must be made... in which the quantum of risk is placed in one scale and the sacrifice involved in the measures necessary for averting the risk (whether in time, trouble or money) is placed in the other and that, if it be shown that there is a great disproportion between them – the risk being insignificant in relation to the sacrifice – the person upon whom the obligation is imposed discharges the onus which is upon him".

The meaning of 'reasonably practicable' has evolved with case law since 1949 in a number of jurisdictions. In Australia a leading case was *Turner v State of South Australia*[8] before the

---

[6] See the following references for further details about 'reasonably practicable' - Safe Work Australia, 2011, *Interpretative Guideline "The Meaning of Reasonably Practicable"* and Johnstone, R. and Tooma, M., 2012, *"Work Health and Safety Regulation in Australia – The Model Act"*, The Federation Press.

[7] [1949] All ER 743 CA.

[8] [1982] 42 ALR669.

High Court of Australia. Chief Justice Gibbs, in his judgment, stated:

"Where it is possible to guard against a foreseeable risk, which, though perhaps not great, nevertheless cannot be called remote or fanciful, by adopting a means, which involves little difficulty or expense, the failure to adopt such means will in general be negligent".

Justice Gaudron of the High Court of Australia went further in *Slivak v Lurgi*[9], setting out three general propositions on the meaning of "reasonably practicable":

- "Reasonably practicable" means something narrower than "physically possible" or "feasible".
- What is "reasonably practicable" is to be judged on the basis of what was known at the relevant time.
- To determine what is "reasonably practicable" it is necessary to balance the likelihood of the risk occurring against the cost, time and trouble necessary to avert that risk.

The High Court of Australia also stated in *Kirk v Industrial Relations Commission*[10] that:

"The measures which must be taken are those which are reasonably practicable. The term is not defined in the OH&S Act, but it may often involve a common sense assessment".

In light of these cases, "reasonably practicable" is essentially an "equity" (ie. fairness) and "utility" (ie. usefulness) based criteria for making decisions about safety-related risks.

Reasonably practicable promotes "equity" because it supports the premise that all individuals have unconditional rights to certain levels of protection. Worker safety must be ensured so far as is reasonably practicable. The primary duty applies

---

[9] [2001] HCA 6.
[10] [2010] HCA 1.

equally to all PCBUs and no PCBU owes a lesser duty than any other.

Reasonably practicable also promotes "utility" because it implies an approach to safety management that strives to find the most suitable and effective risk controls, and allows for certain risk controls to be discounted on the basis of non-effectiveness, unsuitability or gross disproportionality of cost to risk benefit.

'Reasonably Practicable' has since been defined in the Model Act (Section 18) as:

"...that which is, or was at a particular time, reasonably able to be done to ensure health and safety, taking into account and weighing up all relevant matters, including:

a. The likelihood of the hazard or the risk concerned occurring.

b. The degree of harm that might result from the hazard or the risk.

c. What the person concerned knows, or ought reasonably to know, about the hazard or risk, and ways of eliminating or minimising the risk.

d. The availability and suitability of ways to eliminate or minimise the risk, and

e. After assessing the extent of the risk and the available ways of eliminating or minimising the risk, the cost associated with available ways of eliminating or minimising the risk, including whether the cost is grossly disproportionate to the risk".

What is 'reasonably practicable' is determined objectively, in law. This means that a duty-holder must meet the standard of behaviour expected of a reasonable person in the duty-holder's position and who is required to comply with the same duty.

There are two elements to what is 'reasonably practicable'. A duty-holder must first consider *what is able to be done* - that is, what is possible in the circumstances for ensuring health and safety. They must then consider whether it is *reasonable, in the circumstances* to do all that is possible.

It is possible that, in certain circumstances, measures that are able to be implemented, may not be "reasonable" to implement. It may be the case that some things that are able to be done are unsuitable because they are not effective against the risk in the particular circumstances, or they may introduce more risk than they mitigate, or they counteract other things put in place against the risk. This aspect is addressed later in the book through an exploration of the concept of "control effectiveness".

Figure 2.2 shows what might occur were it not for the "reasonable" part of the "reasonably practicable" calculus.

Figure 2.2: A Race-horse after WH&S

A PCBU must determine what is able to be done, and whether it is <u>reasonable</u> in the circumstances to do all of it.

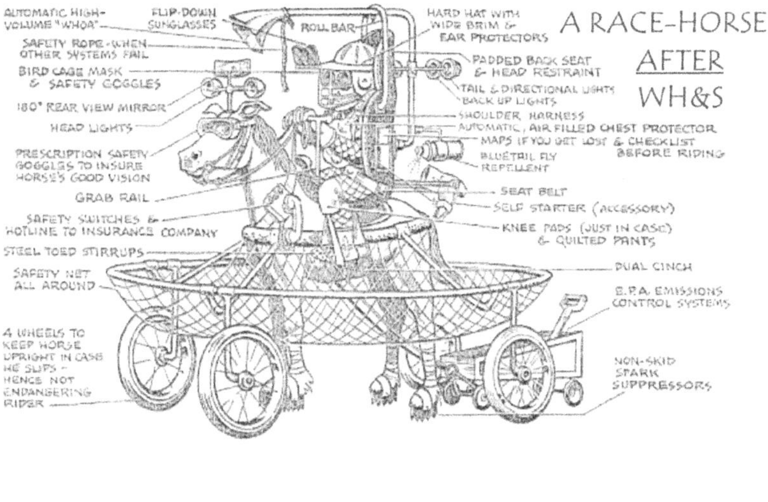

Nevertheless, all of the five (5) factors listed in Section 18 of the Model Act must be considered. No single factor alone or sub-set of factors can determine what is (or is not) reasonably practicable.

With respect to the 'likelihood of the hazard or risk', the Model Act intends that the greater the likelihood of a risk eventuating, the greater the significance this will play when weighing up all matters in determining what is reasonably practicable. If harm is more likely to occur, then it may be reasonable to expect more to be done to eliminate or minimise the risk[11].

With respect to the 'degree of harm that might result', the Model Act intends that the greater the degree of harm that could result from the hazard or risk, the greater the significance this will play when weighing up all matters in determining what is reasonably practicable. If the degree of harm could be significant, then it may be reasonable to expect more to be done to eliminate or minimise the risk[12].

Whilst likelihood and the degree of harm are clearly included in the determination of what is reasonably practicable, the use of their product, commonly known as 'risk level', may prove problematic. Risk level, of itself, usually provides little or no resolution of the component likelihood or degree of harm factors from which it was derived. Furthermore, an estimate of the likelihood of an event may be subject to considerable uncertainty, particularly when analysing high consequence events which have not occurred in the past, but which nonetheless are seen to have the potential to occur.

---

[11] See Safe Work Australia, 2011, *Interpretative Guideline "The Meaning of Reasonably Practicable"*.

[12] See Safe Work Australia, 2011, *Interpretative Guideline "The Meaning of Reasonably Practicable"*.

For example a very high consequence but very low likelihood event may not be classed as a high level 'risk' when the consequence and likelihood parameters are combined[13]. If risk level is used in the determination of 'reasonably practicable', instead of its component factors, then the fact that the event has very high consequences may be overlooked and have less significance to the selection of risk controls than it should have had. It may have been the case that proper consideration of the very high consequences resulted in a decision that more needed to be done about the risk, or that the activity giving rise to the risk not being undertaken. This failing has often been observed by commentators following large industrial disasters, including the BP Texas City refinery accident and the Esso Longford 'disaster'[14].

---

[13] For example, this may happen if a risk matrix is used to evaluate the risk level. The risk levels resulting from each combination of the likelihood and consequence factors depend on how the risk matrix has been constructed. It is often the case that very high consequence / very low likelihood events are classed as "moderate" or "medium" risks.

[14] *"Their view is that our system was insufficient to systematically and continuously identify, reduce and manage risk, in particular to identify low probability high consequence risks."* Lord Browne's (CEO of BP plc.) remarks at a press conference on 16th January 2007 in response to the release of the Baker Panel Report into the Texas City refinery accident.

"Safety rather than engineering has dominated the lessons from the disaster....The safety focus has resulted in risk management being aimed at reducing high-frequency, low consequence personal injuries. This is different from an engineering focus that is essential to reducing low-frequency, high-consequence catastrophes. A balance needs to be struck that does not neglect either end of the risk spectrum". From McNicol, James., 2001, *"Have Australia's Major Hazard Facilities Learned from the Longford Disaster?"*, IEAust Public Policy Unit, p. 24.

In light of this, when making decisions about how to control safety risks, the use of 'risk level' should be constrained as follows:

- The level of risk which remains when all of the selected risk controls for a particular hazard / risk are put into place cannot be used to justify a position that nothing else needs to be done, or that there are no further 'reasonably practicable' risk controls, nor can it be used to justify the acceptance of the risk.
- In particular, for low level risks, the level of risk which would exist without having any risk controls in place for a particular hazard cannot be used to justify a position that it is not necessary, or reasonably practicable for any risk control to be implemented for a particular hazard.

Indeed, many PCBUs may conclude from this that 'risk level' is not particularly relevant to making decisions about the control of, or the acceptability / tolerability of safety risks.

Aspects of the PCBU's financial position (particularly their incapacity to pay for a risk control) cannot be taken into account in determining what is reasonably practicable. If a particular risk control cannot be afforded by a PCBU this could mean that the activity giving rise to the hazard or risk would need to cease or not be undertaken[15].

With respect to 'gross disproportionality', whilst this is also a factor to be 'weighed-up', PCBUs should understand that it may be difficult for a PCBU to justify to a court why it did not implement a 'practicable' risk control on the basis of cost, where the threshold for gross disproportionality has not been

[15] See Safe Work Australia, 2011, *Interpretative Guideline "The Meaning of Reasonably Practicable"*.

established [16]. If the degree of harm is significant (ie. serious injury or death) a "gross disproportionality" argument discounting the use of a known, available and suitable risk control is particularly unlikely to be justifiable.

The concept of "reasonably practicable" has been embodied in the terms SFAIRP, ALARP and ALARA, which are defined as:

SFAIRP – So Far As Is Reasonably Practicable

ALARP – As Low As Reasonably Practicable

ALARA – As Low As Reasonably Achievable

Importantly, the Model Act specifically uses the words "so far as is reasonably practicable" (SFAIRP) and no other combination of words.

With respect to SFAIRP, the PCBU must ensure "so far as reasonably practicable" the health and safety of workers.

With respect to ALARP (and ALARA), the ALARP principle states that: "residual risk shall be as low as reasonably practicable (achievable)".

In literal terms, SFAIRP is associated with the stating of a desired outcome – the assurance of worker health and safety. ALARP (and ALARA) are associated with stating a desired level of risk – the level of risk which is not reasonably practicable to reduce any further. Various publications make it clear that when correctly applied, SFAIRP and ALARP, whilst semantically different, actually aim to achieve the same practical outcome.

---

[16] See Johnstone, R. and Tooma, M., 2012, *"Work Health and Safety Regulation in Australia – The Model Act"*, The Federation Press, p.74.

For instance, the United Kingdom Health and Safety Executive website[17] states:

"In terms of what they require of duty-holders, HSE considers that duties to ensure health and safety so far as is reasonably practicable ("SFAIRP") and duties to reduce risks as low as is reasonably practicable ("ALARP") call for the same set of tests to be applied."

The application of SFAIRP and ALARP is illustrated in Figure 2.3.

---

[17] See http://www.hse.gov.uk/risk/theory/alarp1.htm#P14_1686

Figure 2.3: SFAIRP V's ALARP

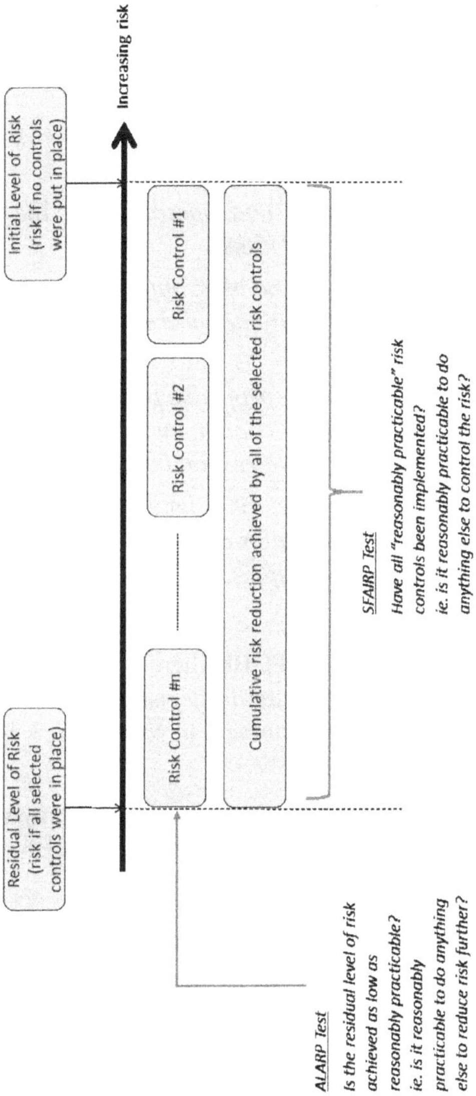

The Duty of an Officer of a PCBU[18]

Under the Model Act (Section 27) an officer of a PCBU must exercise "due diligence" to ensure the PCBU complies with its health and safety duties. This "due diligence" duty requires an officer of a PCBU to take reasonable steps to:

a. *acquire and keep up-to-date knowledge on work health and safety matters,*
b. *understand the nature and operations of the work and associated hazards and risks,*
c. *ensure the PCBU has, and uses, appropriate resources and processes to eliminate or minimise risks to work health and safety,*
d. *ensure the PCBU has appropriate processes to receive and consider information about work-related incidents, hazards and risks, and to respond in a timely manner,*
e. *ensure the PCBU has, and implements, processes for complying with their duties and obligations, and*
f. *Verify the provision and use of the relevant resources and processes.*

The duty of the officer of a PCBU therefore relates to strategic, structural, policy and resourcing decisions - that is, how the business or undertaking is run and how safety risk is managed in general.

This is a fundamentally different duty to the primary duty of the PCBU (described in a previous section), which is focussed

---

[18] See the following references for further details about officers of PCBUs - Safe Work Australia, 2011, *Interpretative Guideline "The Health and Safety Duty of an Officer Under Section 27"*, Workcover NSW, 2011, *Fact Sheet "PCBUs, Workers and Officers"* and Johnstone, R. and Tooma, M., 2012, *"Work Health and Safety Regulation in Australia – The Model Act"*, The Federation Press.

on the adequacy of the actual measures taken against the particular risks posed to the workers and others. The duty of the officer/s of a PCBU is shown diagrammatically in Figure 2.4 (within the dotted line).

Figure 2.4: Duty of an Officer of a PCBU

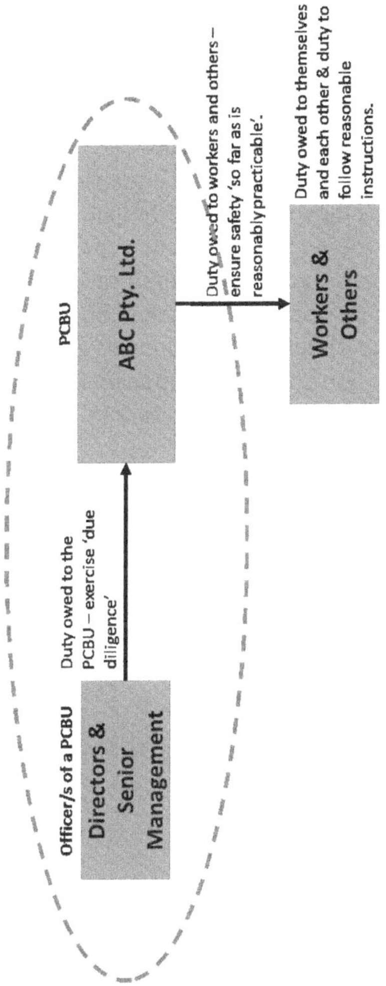

Failure to comply with this duty is a criminal offence and an officer of a PCBU can be prosecuted for not properly exercising 'due diligence'. The exception to this rule is a volunteer who is an officer of a PCBU. Whilst they have a duty to exercise due diligence, they cannot be prosecuted for failing to comply with that duty. A volunteer officer of a PCBU will still be expected to exercise 'due diligence' and a regulator can compel compliance with that duty by issuing an improvement notice.

As previously stated, a worker, including a member of a profession (e.g. an engineer, lawyer, accountant, teacher, doctor, etc…) or a manager in an organisation is unlikely to be an officer unless they make or participate in making decisions that affect the whole, or a substantial part, of the business or undertaking. Therefore, only a small proportion of the people working within an organisation would be classed as officers of a PCBU (eg. the directors, the company secretary, the general manager, the design/engineering manager for designer PCBUs, construction manager for construction PCBUs etc…).

Chapter 3 - Implications of the Model Act and Regulations

Foreseeable Hazards

The Model Regulation (Section 34) requires a PCBU to identify 'reasonably foreseeable' hazards that could give rise to health and safety risks. If a hazard is 'reasonably foreseeable', a PCBU will be required to take measures to control it. Therefore, if it is not reasonable to foresee a particular hazard then a PCBU will not be required to take measures to control it.

'Foreseeability' has a particular meaning in law - the facility to perceive, know in advance, or reasonably anticipate that damage or injury will probably ensue from acts or omissions[19]. That is, the law considers whether a person of ordinary intelligence and circumspection, should reasonably have foreseen that his or her act would imperil others. This consideration can include the event that transpired or similar occurrences. It is not necessary to give regard to what the person involved in the actual event thought would happen in regard to the actual event or the manner of causation of injuries.

To understand the operation of reasonable foreseeability in law requires an understanding of the area of the tort law known as 'negligence'[20].

In tort law, the negligence (and liability) of a defendant who has a 'duty of care'[21] to a plaintiff arises from 'cause-in-fact'

---

[19] See http://legal-dictionary.thefreedictionary.com/Foreseeability
[20] The tort of negligence only deals with harm caused by carelessness, not intentional harm.
[21] A 'duty of care' refers to the idea that a standard of reasonable care is required where an act or omission could foreseeably harm others.

and 'proximate cause' with respect to the injury or harm suffered by a plaintiff. Both causes must be proven[22].

The proof of 'cause-in-fact' requires that the defendant's conduct was the actual cause of the plaintiff's injuries. Courts use the "but for" or "sine qua non" rule to determine the 'cause-in-fact'. This rule considers whether the injury would not have occurred "but for" the defendant's act or omission. For example: "but for" the defendant running the red light, the collision would not have occurred – the 'cause-in-fact' was the defendant running the red light.

The proof of 'proximate cause' requires that the injuries or harm suffered by the plaintiff must have been a foreseeable result of the defendant's conduct. A 'proximate cause' is an act or omission from which an injury results as a natural, direct, uninterrupted consequence and without which the injury would not occur. Reasonable foreseeability is used to determine proximate cause. The 'foreseeability test' basically asks whether the person causing the injury should have reasonably foreseen the consequences that would result from their conduct. For example, it is reasonably foreseeable that if a worker drops a brick on another worker's head it could cause them a blunt-force injury. The 'foreseeability test' requires that both the plaintiff (ie. the injured or harmed person) and the type of

---

[22] This section has drawn upon material located at the following webpages:
http://sixthformlaw.info/01_modules/other_material/tort/1_duty/3_duty_foreseeable.htm
http://www.unistudyguides.com/wiki/Reasonable_foreseeability
http://legal-dictionary.thefreedictionary.com
http://www.lexisnexis.com/uk/lexispsl/personalinjury/synopsis/448:449/Establishing-legal-liability/Proving-negligence-or-breach-of-statutory-duty

injury or harm suffered by the plaintiff must have been foreseeable.

Proximate causes automatically produce foreseeable consequences except when there is intervention by an independent or unforeseeable cause (i.e. a separate act or omission that breaks the direct connection between the defendant's act or omission and the injury). A defendant cannot be held liable for injuries that are unpredictable and do not have a reasonable link to their conduct.

For example, suppose that a speeding motorist collided with a truck. Upon impact, the truck exploded and the shock wave caused a traffic light to collapse one kilometre away, which injured a by-stander. The speeding driver would likely not be held liable for the by-stander's injuries as it not reasonably foreseeable that the collision would affect a by-stander one kilometre away and it is not reasonably foreseeable that such injuries would have resulted from the collision. However, the speeding driver may be liable for injuries caused to the truck driver and others at the scene of the crash.

'Reasonably foreseeable' carries a narrower definition than 'foreseeable', but there is substantial case law on the matter, including:

In *Donoghue v Stevenson*[23] Lord Atkins established the 'neighbour principle', which is *"you must take reasonable care to avoid acts or omissions which you can reasonably foresee would be likely to injure your neighbour."*

Bolton v Stone[24] distinguished between cases where the possibility of the risk was so far-fetched that no reasonable man would have done anything and cases in which the risk was material. It established that harm is reasonably foreseeable if it

---

[23] [1932] UKHL 100.
[24] [1951] AC 850.

isn't *"thought to be physically impossible or because the possibility of its happening would have been regarded as so fantastic or farfetched that no reasonable man would have paid any attention to it"*.

The test of what is 'reasonably foreseeable' was also considered in *Wyong Shire Council v Shirt*[25]. It established that a risk does not have to be probable or likely to occur to be reasonably foreseeable. An unlikely risk can also be reasonably foreseeable, but not one which is "far-fetched or fanciful".

The role of the "but for" and "foreseeability" tests in establishing the "cause-in-fact" and "proximate cause" and their ultimate role in proving negligence are illustrated in Figure 3.1.

---

[25] [1980] 146 CLR 40.

Figure 3.1: Foreseeability and Negligence

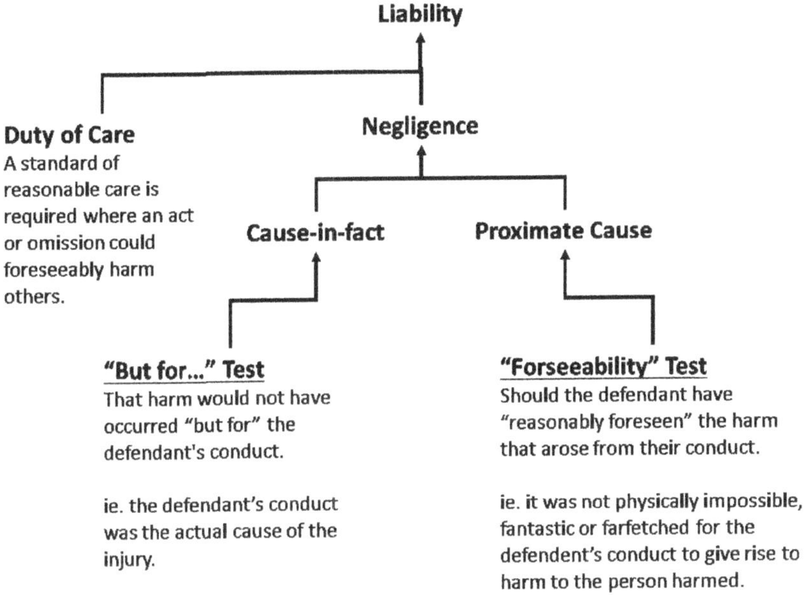

If a hazard is not reasonably foreseeable in law, then a PCBU will not be required to take measures to control it.

For example, Mike works in John's shop. Whilst working in the shop one day, Mike decides to climb onto a bench and jump across to another bench. He doesn't land properly, falls and seriously injures his eye on a drinking glass that was left sitting on the second bench by another person.

John owes a duty of care to Mike because Mike is a worker in his shop. It is reasonably foreseeable that an act or omission by John could lead to Mike being harmed. However, it is not reasonably foreseeable that a risk is created by someone leaving a drinking glass on a bench. Leaving a glass on the bench does not give rise to the possibility that someone will fall on it whilst jumping between benches. Thus, the 'foreseeability test' will not be satisfied. John therefore does not have an obligation to control the specific risk whereby a worker jumps from bench to bench and injures themselves on a drinking glass left on one of the benches[26]. However, he may have the obligation to control the more general risk whereby a worker climbs onto a bench and falls off it onto the floor and is injured.

Within the context of safety management, the identification of 'reasonably foreseeable' hazards must therefore be focussed on identifying all hazards, with the exception of those that are physically impossible, or those whose probability of occurrence is so far-fetched or fanciful that no reasonable person would pay them attention.

Much has been published on the subject of risk identification techniques. The International Electro-technical Commission (IEC) has published an international standard, IEC31010 *Risk Management – Risk Assessment Techniques*. Standards Australia and Standards New Zealand have also published a guideline, *SA/SNZ HB89 Risk Management – Guidelines on Risk Assessment Techniques*. The content of both documents has much in common[27].

These documents provide guidance on the selection and use of various risk assessment techniques, including techniques for identification of hazards / risks.

---

[26] See the following reference for further information: http://www.unistudyguides.com/wiki/Reasonable_foreseeability

[27] This is because a number of members of the IEC working group which drafted IEC31010 were also on the SA/SNZ working group which drafted HB89. IEC31010 was originally intended to replace HB89, but SA/SNZ chose not to adopt IEC31010 because some aspects of the final draft were not agreed upon by the Australian and New Zealand working group members. SA/SNZ chose to pursue its own version of the subject matter, which was later published as a new version of HB89.

SA/SNZ HB89 states that techniques should be selected, based on applicable factors such as the following:

- The purpose of the study
- The needs of decision-makers and other stakeholders
- The type and range of risks being analysed
- The potential magnitude of the consequences
- The degree of expertise, human and other resources needed
- Constraints on time and other resources and opportunity costs
- The complexity of the application
- The nature and extent of uncertainty in available data and information
- The need for modification/updating of the risk assessment
- Any regulatory and contractual requirements will affect technique choice

Further to this, Dr. Tilman Rasche of the Queensland Department of Employment, Economic Development and Innovation (DEEDI) has described pictorially how the various risk assessment techniques might be applied via his "Meaning of Life" chart, as shown in Figure 3.2.

Figure 3.2: Rasche's "The Meaning of Life"

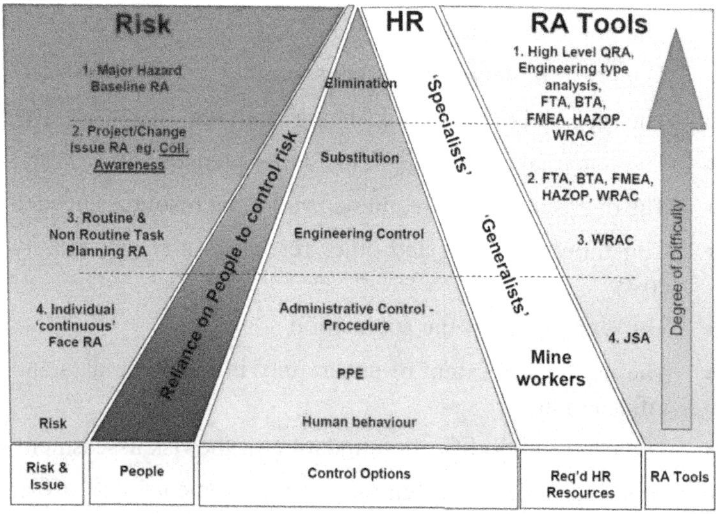

Figure 3.2 shows the relationship between types of risks, risk control options, risk analysis tools and risk practitioner skill requirements. It suggests that as the consequence or level of risk increases the degree of rigour and sophistication required of the risk assessment tools should also increase.

Workplaces involving simple, routine or straight-forward work tasks or using simple or routinely-used technologies (eg. servicing a vehicle, painting a house, welding, operating a chainsaw, general electrical work) would therefore only require simple or informal techniques such as Job Safety Analysis (JSA), Safe Work Method Statement (SWMS) or Workplace Risk Assessment and Control (WRAC) to identify hazards / risks.

Where there is complex technology involved, or work tasks are complex or non-routine or involve potentially high consequences (eg. ventilating a coal mine, operating a rail network or a process plant, designing or modifying the design of a vehicle's brakes, automating a production line), the use of well-tried techniques to identify the hazards / risks is strongly recommended. Techniques such as the HAZOP and FMEA methods[28] use formal, facilitated, systematic, multi-disciplinary team approaches (see Figure 3.3). These techniques increase the chance that reasonably foreseeable hazards are identified.

---

[28] HAZOP is the abbreviation of Hazard and Operability Study. FMEA is the abbreviation of Failure Modes and Effects Analysis. See *SA/SNZ HB89 Risk Management – Guidelines on Risk Assessment Techniques* or *IEC31010 Risk Management – Risk Assessment Techniques.*

Figure 3.3: Formal Risk Assessment

Formal, systematic, multi-disciplinary team approaches increase the chance that foreseeable hazards/risks are identified.

However, formal techniques provide no guarantee that all foreseeable hazards will be identified. They should therefore be supplemented by other methods of hazard / risk identification, including:

- Reviews of relevant standards, codes of practice, regulations and industry guidelines.
- Reviews of safety alerts, bulletins and incident investigation reports.
- Surveys of industry, company and site incident / accident data.
- Informal / ad-hoc hazard risk identification (eg. via hazard issues raised by workers).
- Information obtained via industry working groups, round-tables, seminars and discussions with peers and regulators.

Hazard Elimination and Risk Control

Sections 35 and 36 of the Model Regulations, require that a PCBU must eliminate hazards, "so far as is reasonably practicable", or if this cannot be done, they must minimise risk "so far as is reasonably practicable" according to a hierarchy of control measures[29], as shown in Figure 3.4.

Figure 3.4: Hierarchy of Control Measures

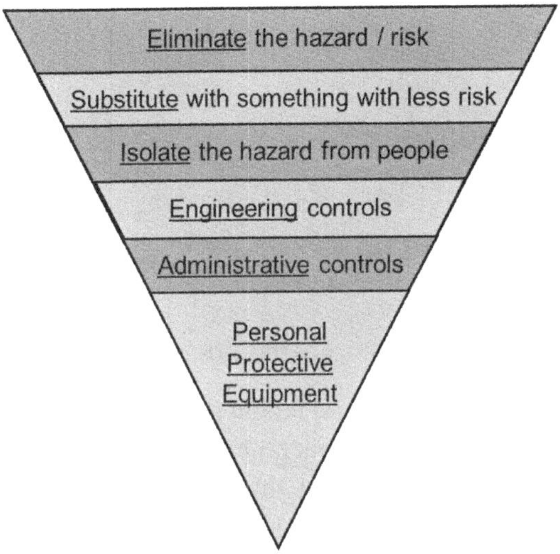

Referring again to Figure 3.2 (The Meaning of Life), this diagram suggests that as the consequence or level of risk increases, the degree to which a risk should be controlled by measures at the top of the hierarchy of risk controls should also increase. These tend to have the highest degree of control effectiveness (see Chapter 4 for further details on the assessment of control effectiveness).

[29] Also commonly referred to as a "hierarchy of risk controls".

The practical operation of the hierarchy of risk controls is illustrated in Figure 3.5. In this example a large air-conditioning unit has been placed above the door to an office.

Figure 3.5: Elimination V's Minimisation

Risk elimination must come before risk minimisation. In other words - move the air-conditioner to ground level and away from the door!

The placement of a mass at height gives rise to potential energy. If the mass were to fall, its potential energy would convert to kinetic energy as it fell. If the falling mass were to strike a person, some or all of the energy may be transferred to the person and cause them harm. A safety hazard is created with the placement of the air-conditioning unit above the door and a risk exists if there is the potential for the unit to fall onto a person. Whilst the likelihood of the unit falling may be conceivably low, it is not a far-fetched or fanciful proposition. Such an event might be caused by corrosion of the mounting brackets or fixings, or by some external event such as an earthquake.

When viewed through the prism of the hierarchy of risk controls, the most effective risk control is to eliminate the hazard, which in this case means to remove the potential energy created by having it mounted above the door where people often pass. Better placement of the air-conditioning unit at a lower height and away from the door would be a more effective control than an engineering control such as stronger, corrosion-resistant mounts or an administrative control such as a maintenance procedure for the mounts or a site rule to use another door.

As previously stated, Figure 3.2 (The Meaning of Life) suggests that as the consequence or level of risk increases, the degree of rigour and sophistication required of the risk assessment tools should also increase.

The use of risk analysis methodologies that link the causes of identified hazards / risks with their consequences and risk controls[30] is strongly recommended where there is complex technology involved, or work tasks are complex or non-routine or involve potentially high consequences.

Furthermore, the consideration of past events, safety alerts and bulletins, investigations reports on incidents and accidents, guidelines, standards and codes of practice can greatly assist in identifying controls for hazards / risks.

---

[30]Such as the 'Bowtie' method, Fault Tree Analysis, Event Tree Analysis or Cause-Consequence Analysis. See *SA/SNZ HB89 Risk Management – Guidelines on Risk Assessment Techniques* or *IEC31010 Risk Management – Risk Assessment Techniques.*

Tolerating or Accepting Risk

'Tolerable risk' (and acceptable risk) refers to a willingness by society as a whole to live with a risk so as to secure certain benefits, in the confidence that the risk is one that is worth taking and that it is being properly controlled. It does not imply that every individual is willing to live with a risk (ie. that everyone would agree without reservation to take the risk or have it imposed on them)[31].

Tolerable risk (and acceptable risk) is not an objective measure, because it is influenced by the social and cultural context that an organisation works within. Therefore, notions of what is a tolerable risk may vary between organisations, industries and countries. For example one organisation may define its tolerable risk level as being the likelihood of a fatality of less than 0.001 per year, whilst another organisation in the same industry may define it as less than 0.0001 per year. The difference is significant and its impact could mean many lives are saved or lost.

Therefore, tolerable risk (and acceptable risk) is a subjective test for determining acceptability of health and safety risks to workers and this is inconsistent with the objective test of "so far as is reasonably practicable" in the Model Act.

Nevertheless, a number of current Australian and International standards, codes of practice and guidelines use the notion of tolerable risk as a basis for determining the reliability, integrity, strength or robustness of risk controls for particular hazards.

Examples of this are:

---

[31] United Kingdom Health and Safety Executive, 2001, *"Reducing Risk, Protecting People - HSE's decision-making process"*, HSE Books, p. 3.

- AS/NZS ISO31000 risk management standard and its companion handbook HB436[32], which allow for (but do not mandate) the acceptance of a risk on the basis of residual risk level being lower than a tolerable risk level criteria.
- AS(IEC)61508[33] and associated functional safety standards use a notion of tolerable risk level to determine the reliability (safety integrity) requirements of risk controls based on electrical and electronic technologies.
- AS(IEC)60079 series of standards for hazardous areas which determine the requirements for explosion protection measures for a hazardous area on the basis of the likelihood of a flammable gas or dust becoming present in that area.
- Energy Networks Association of Australia (ENA) Guideline EG-0 power system earthing guide uses a notion of tolerable risk level to determine the requirements for electrical earthing of installations.

These documents and other documents of this nature are important to the specification and implementation of certain types of risk controls, but clearly, the requirements of the Model Act must take priority if a management system for safety-related risk is to comply with the Model Act. Organisations must therefore ensure that the criteria for acceptance of work safety risks is based on the achievement of the primary duty as a PCBU, that is, a duty-holder must first consider *what can be done* - that is, what is possible in the circumstances for ensuring health and safety. They must then

---

[32] SA/SNZ HB436 Risk Management Guidelines – Companion to AS/NZS ISO31000.

[33] Some Australian standards are referred to throughout this book as "AS(IEC)...". This was done for the sole purpose of expressing their origin and equivalence to the IEC standards.

consider whether it is *reasonable, in the circumstances* to do all that is possible.

Specifically, this means that:

- The level of risk which remains when all of the selected risk controls for a particular hazard are put into place cannot be used to justify a position that nothing else needs to be done, or that there are no further 'reasonably practicable' risk controls, nor can it be used to justify the acceptance of the risk.
- In particular, for low level risks, the level of risk which would exist without having any risk controls in place for a particular hazard cannot be used to justify a position that it is not necessary, or reasonably practicable for any risk control to be implemented for a particular hazard.
- It may be difficult for a PCBU to justify not implementing a risk control on the basis of cost. If the degree of harm is significant (ie. serious injury or death) a "gross disproportionality" argument discounting the use of a known, available and suitable risk control is unlikely to be justifiable[34].
- Aspects of a PCBUs financial position (particularly their incapacity to pay for a risk control) cannot be taken into account in determining what is reasonably practicable. In order to comply this could mean that the activity giving rise to the hazard or risk would need to cease or not be undertaken.

In those instances where tolerable risk approaches are used (whether mandatory or voluntary), it is recommended that

---

[34] See Johnstone, R. and Tooma, M., 2012, *"Work Health and Safety Regulation in Australia – The Model Act"*, The Federation Press, p.74.

PCBUs obtain a level of confidence that their primary duty is met by the selected regime of risk controls for a hazard / risk before applying a tolerable risk approach to determine the reliability, integrity, strength or robustness requirements of particular risk controls within that regime.

An example of the operation of these principles is given below and in Figure 3.6.

A known and available risk control against the risk of people falling from a railway platform, or being struck by a train at a railway platform is a technology known as "platform screen doors". These are electronically controlled doors that form a barrier between the platform and the railway track. They are designed to open and close with the arrival and departure of trains at the platform. Platform screen doors control a hazard / risk with a potential for significant harm (ie. serious injury or death) to workers or others who use a railway system. Many railway operators world-wide have installed them on new or existing railway platforms and their effectiveness at preventing harm is well documented.

Under the Model Act a PCBU building or upgrading a railway station cannot justify a position to not install platform screen doors on the basis that the level of risk is already sufficiently low due to other measures taken[35]. The PCBU must determine whether it is reasonable to install platform screen doors, which are a known and available risk control. It will be difficult for the PCBU to justify not installing them when many other railway operators have already done so in similar circumstances and validated their effectiveness.

---

[35] For example - yellow lines painted near the platform edge, CCTV cameras or regular public announcements warning passengers on the platform to stay away from the platform edge.

Furthermore, it will be difficult for the PCBU to justify not installing them on the basis of their cost being "grossly disproportionate" to the safety benefits because the potential degree of harm is significant. The PCBUs financial position and its ability to afford platform screen doors cannot be taken into account either.

In this case a tolerable risk approach (such as that provided by AS(IEC)61508 could be taken to determine the appropriate level of reliability that should be designed into the electronic system which controls the platform screen doors.

Figure 3.6: Think Control, Not Level of Risk !

A foreseeable hazard – falling from a railway platform and/or being struck by a train.

A 'reasonably practicable' risk control – "platform screen doors".

Due Diligence Vs Reasonably Practicable

In the Model Act, "due diligence" and "reasonably practicable" are concerned with different duties. PCBUs must be aware that a 'due diligence' approach by their officers does not release it from its duty to execute a 'reasonably practicable' approach with respect to its primary duty as a PCBU. Furthermore, officers of PCBU's must be aware that a 'reasonably practicable' approach by the PCBU does not release them from executing the 'due diligence' approach with respect to their duty as an officer of a PCBU. Both duties must be executed by the relevant duty holders.

Due diligence is concerned with how far the officers of a PCBU have ensured that the PCBU has met its overall duties and obligations (ie. how the workplace is run). The Model Act recognises that it is possible for officers of a PCBU to meet their due diligence duties without the PCBU having implemented all reasonably practicable risk controls for all foreseeable hazards / risks. Therefore, a PCBU may be charged with an offence under the Model Act whether or not the officers of that PCBU have been found guilty of an offence under the Model Act.

'Reasonably practicable' relates to whether a PCBU has met its primary duties and obligations with respect to its workplace safety (ie. whether all reasonably practicable risk controls have been implemented). The Model Act recognises that it is possible for a PCBU to have met its primary duties without the officers having met their due diligence duties. Therefore, an officer of a PCBU may be charged with an offence under the Model Act whether or not the PCBU has been found guilty of an offence under the Model Act.

The Relevance of 'Safety Cases'

A 'safety case' is a structured, pre-emptive argument, supported by evidence, that a system is 'acceptably' safe. Sometimes a safety case is required by regulations governing some industries, particularly those where there has been the potential for high consequence accidents (eg. major hazard facilities, railways, off-shore oil/gas facilities etc…).

Safety cases, if focussed on demonstrating SFAIRP with respect to hazards / risks, are therefore an effective means for a PCBU to document and communicate to stakeholders that the health and safety of workers has been ensured so far as is reasonably practicable. Such safety case documents may also be useful in demonstrating that a PCBU's primary duty was properly discharged in the event of an incident or accident.

Whilst a safety case may delve into the Occupational Health and Safety Management System (OH&S MS) of an organisation, it would not normally be expected to demonstrate that the officers of the PCBU have exercised their due diligence duty. A safety case demonstrates the assurance of safety through a systematic argument that each identified risk has been adequately controlled. It does not normally venture into the adequacy of the strategic, structural, policy and resourcing aspects of the PCBU - that is, how the business or undertaking is run.

A structured Safety Case argument and the supporting evidence may be summarised using Goal Structuring Notation (GSN) as shown in Figure 3.7.

Figure 3.7: Safety Case Argument

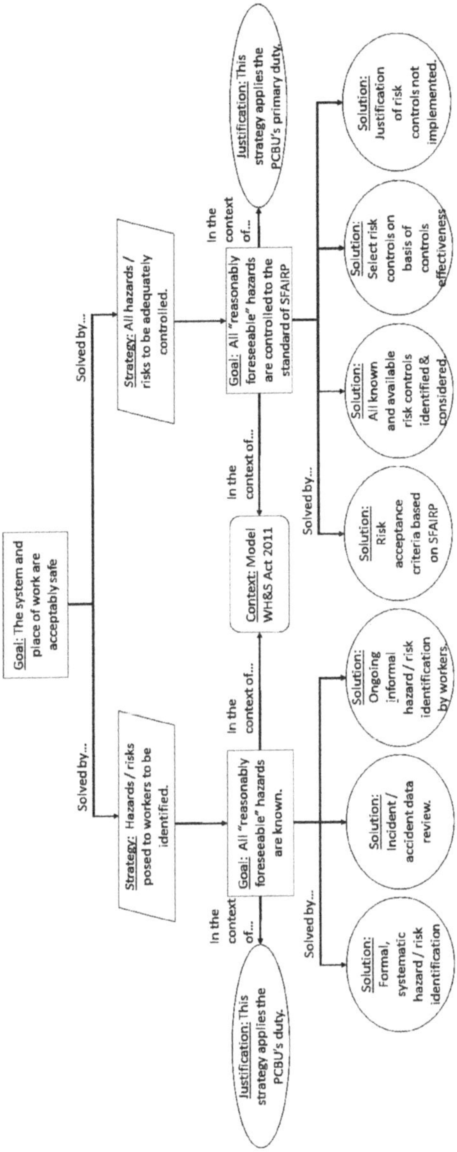

Chapter 4 - Approaches to Compliance

Risk Management Standards

AS/NZS ISO31000 and SA/SNZ HB436

A key standard for risk management in an organisation is *AS/NZS ISO31000 Risk Management – Principles and Guidelines*. Its companion guideline is *SA/SNZ HB436 Risk Management Guidelines – Companion to AS/NZS ISO31000.* These documents provide the principles and processes and a framework for establishing and maintaining a risk management system - which can be applied to a variety of risk contexts, such as safety, environmental, financial etc...

The overall concept of the AS/NZS ISO31000 risk management process is shown in Figure 4.1[36]. Whilst not depicted in Figure 4.1, risk evaluation and risk treatment are actually a cyclical process within the main risk management process.

It is clearly the intention of the standard that the generic risk management process of Figure 4.1 may be tailored to the specific risk context at hand. There is sufficient flexibility in AS/NZS ISO31000 for the requirements of the Model Act to be incorporated into the generic risk management process to make it consistent with the requirements of the Model Act.

Indeed, Clause 5.3 of AS/NZS ISO31000 establishes a need for an organisation to 'establish the context' with respect to its objectives, internal and external parameters and the scope and risk criteria as the first step of a risk management process. Part of the 'establish the context' step is to establish the 'external context' of the risk management process, that is, the external environment in which the organisation seeks to achieve its objectives. This external context typically includes the

---

[36] See Figure 1 of AS/NZS ISO31000.

physical, social, cultural and legal / regulatory parameters and constraints that relate to what is being assessed.

Figure 4.1: AS/NZS ISO 31000 Process

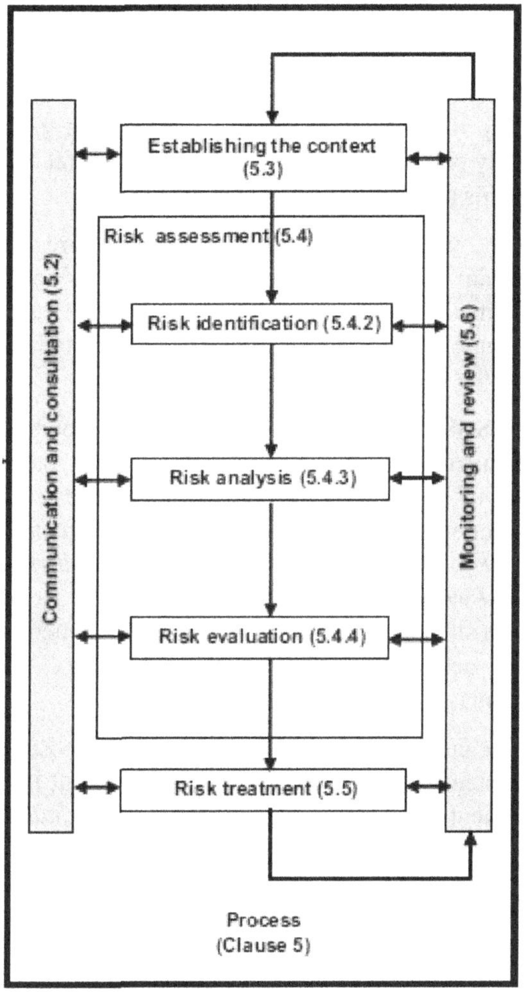

Clearly, the Model Act has implications for the application of the AS/NZS ISO31000 risk management process - particularly at the 'establishing the context', 'risk evaluation' and 'risk treatment' steps.

It is clear that during the 'establishing the context' step of the risk management process, the risk criteria for determining the acceptability of risk must be consistent with the Model Act, and specifically the PCBU's primary duty. The principle of "reasonably practicable" must be the fundamental criterion for deciding this matter.

In this way, when the 'risk evaluation' step is reached, those involved in making decisions about risk controls and acceptability of risk will have a decision criterion that is aligned with the achievement of the PCBU's primary duty in the Model Act.

In AS/NZS ISO31000, 'risk evaluation' is intended to assist decision-making about the treatment of risks, based on the outcomes of the 'risk analysis' step. Clause 5.4.4 of the standard states that risk evaluation *"...involves comparing the level of risk found during the analysis process with risk criteria established when the context was considered"*. However, this clause also states that decisions about risk treatment should be made in accordance with legal, regulatory and other requirements.

However, Clause 5.5.1 (Risk Treatment) of AS/NZS ISO31000 is inconsistent with the Model Act in so far that it states that risk treatment involves *"deciding whether residual risk levels are tolerable"*.

Risk treatment is defined as a cyclical process of:

1. assessing a risk control,
2. deciding whether residual risk levels are tolerable,

3. if not tolerable, generating a new risk control, and
4. assessing the effectiveness of that risk control.

In summary, the emphasis of AS/NZS ISO31000 is risk reduction to a tolerable risk criteria. However, the standard also requires that decisions about risks are to be made in accordance with legal and regulatory requirements. Therefore, an avenue exists for the AS/NZS ISO31000 risk management process to be modified so that it is consistent with the requirements of the Model Act.

The risk evaluation / risk treatment steps should now be taken to involve those aspects already discussed in earlier parts of this book. The following modification to the risk evaluation / risk treatment steps of the AS/NZS ISO31000 process is recommended:

Modified AS/NZS ISO31000 Risk Evaluation / Treatment Steps

1. Determine all known and available risk controls for each hazard / risk.
2. Decide which risk controls are and are not "reasonably practicable" with particular attention to the following:
    a. The suitability of each risk control for the circumstances – taking account of their effectiveness against the risk at hand (see the next section), further risks introduced etc…
    b. Risk elimination measures must be used against a hazard, where "reasonably practicable", before risk reduction measures are considered.
    c. If risk elimination measures are not "reasonably practicable", the hierarchy of risk controls must be used to guide the selection of the most effective risk reduction measures for each hazard.
    d. Justification should be provided for each risk control determined to be not "reasonably practicable" to implement – eg. ineffective in the circumstances, gross disproportionality, net risk increase.
3. Decide which "reasonably practicable" risk controls are to be implemented and not implemented, noting that if a PCBU does not have the capacity to pay for a reasonably practicable risk control, it should cease or not undertake the activity which gives rise to the risk.
4. Determine the reliability, integrity, strength or robustness requirements of any risk controls which require a tolerable risk approach to their specification and design (eg. earthing, electronic safety systems, explosion protection, and structural design). This should also involve a consideration of the level of reliability / integrity / strength / robustness that is reasonably practicable for a designer to achieve, and take account of the specific requirements of regulations or standards for such risk controls.

HB205

The forthcoming Standards Australia handbook *HB205 Managing Safety-related Risk*[37] is intended to provide specific guidance on the application of AS/NZS ISO31000 and SA/SNZ HB436 to safety-related risk. It is the policy of Standards Australia that it is generally unable to provide interpretation of legislation or advice on how to achieve compliance. Therefore, the forthcoming HB205 can only deal with safety-related risk in general terms and cannot provide specific guidance on applying AS/NZS ISO31000 or SA/SNZ HB436 to meet the requirements of the Model Act.

However, at the time of publication of this book the draft of the forthcoming HB205 included clarification of Clause 5.5.1 of AS/NZS ISO31000, which states that part of the risk treatment step of a risk management process involves *"deciding whether residual risk levels are tolerable"*. This clause implies that risk level may be used as a basis for deciding whether a risk may be accepted and nothing more done about reducing its level of risk.

This has been seen by many to be at odds with the concept of "reasonably practicable", where an organisation's health and safety obligations would not be met if an additional risk control was "reasonably practicable" to implement, in spite of a target residual risk level being achieved.

The draft of the forthcoming HB205 clarifies this by emphasising that the "reasonably practicable" test, not notions of a tolerable risk level, must be the criteria for risk acceptance of safety-related risk.

---

[37] The current published version of HB205 is HB205-2004. This handbook is entitled *"OHS Risk Management Handbook"*. An update to HB205 was in draft at the time of publication of this book, with the new title *"Managing Safety-related Risk"*. The author was a member of the Standards Australia working group in 2014/ 2015 that drafted the forthcoming version of HB205.

SA/SNZ HB89 & IEC31010

SA/SNZ HB89 Risk Management – Guidelines on Risk Assessment Techniques and *IEC31010 Risk Management – Risk Assessment Techniques* both provide guidance on the specific techniques that can be used during the 'risk assessment' steps in a risk management process - risk identification, risk analysis and risk evaluation.

SA/SNZ HB89 and IEC31010 are both useful documents for those involved in implementing risk management processes and risk assessments. Unfortunately neither document includes within its scope techniques applicable to the 'establishing the context', 'risk treatment', 'communication and consultation' and 'monitoring and review' steps of the risk management process. For example, neither standard provides techniques for use in the 'risk treatment' step to address the assessment of the effectiveness of a proposed control.

OH&S Management Systems

Occupational Health and Safety Management Systems (OH&S MS) have been seen as integral to successful safety management in organisations for at least two (2) decades. In many organisations OH&S MS is generally placed within the overall business continuous improvement process and aligned with the ISO9001 quality management philosophy, as shown in Figure 4.2.

Figure 4.2: AS/NZS4801 OH&S Model

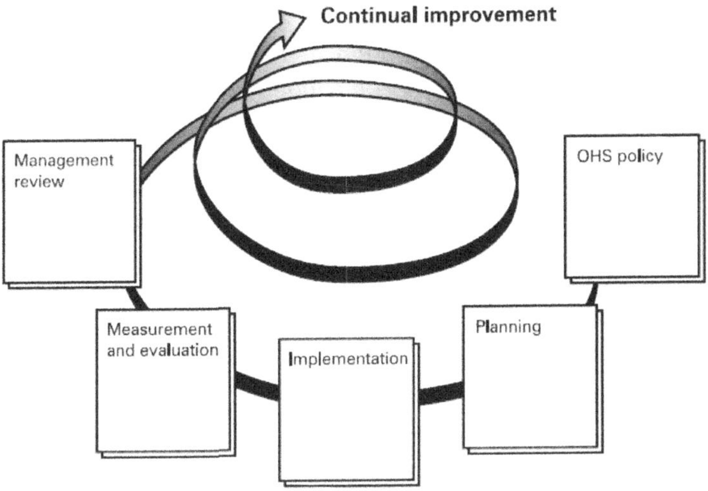

The core elements of an OH&S MS are:

- Clearly defined management responsibility and accountability for OH&S.
- Identification of hazards, associated risks and their management.
- Documentation of procedures associated with the management of hazardous activities.
- Inspections of hazardous plant; and
- Training.

Other elements may be implemented in accordance with organisational needs and priorities.

In Australia (and New Zealand) two (2) standards have been associated with the development and operation of OH&S MS and these are generally applicable to those involved in establishing and maintaining an organisation's OH&S MS.

AS/NZS4801:2001 Occupational Health and Safety Management Systems – Specification with Guidance for Use specifies requirements for an overall Occupational Health and Safety (OH&S) management system. It also establishes an audit framework for use by third parties who may be engaged to conduct independent audits of an organisation's OH&S management systems. The standard is generally applicable to those involved in establishing, maintaining or auditing an organisation's Occupational Health and Safety (OH&S) management system.

AS/NZS4804:2001 Occupational Health and Safety Management Systems –General Guidelines on Principles, Systems and Supporting Techniques provides guidance on the development and implementation of Occupational Health and Safety (OH&S) management systems and principles, and their integration with other management systems. The standard is generally applicable to those involved in establishing and maintaining the organisation's Occupational Health and Safety (OH&S) management system.

The risk management approach recommended in these standards is that provided in AS/NZS4360:2004 (which was superceded by AS/NZS ISO31000 in 2009, however the generic risk management process is virtually identical in both standards). The risk assessment techniques recommended in these standards are those provided in *AS/NZS3931:1998 Risk Analysis of Technological Systems* (which was superceded by SA/SNZ HB89 in 2012/2013, but contains similar content).

In light of previous commentary that an AS/NZS ISO 31000 risk management process can be made to be compatible with the Model Act, it follows that an OH&S MS that is based on such a modified risk management process can also be made to be compatible with the Model Act.

However, OH&S MS (or WH&S MS as they might now be called) appear to have more significant issues to resolve than this.

Systematic analysis of the scientific literature concerning the effectiveness of OH&S MS has shown that despite more than twenty (20) years of world-wide use and development, no conclusive and sound evidence has been obtained that OH&S MS are effective in terms of preventing and reducing the number of occupational accidents and diseases[38].

This finding raises a number of questions worth further contemplation:

- Why are OH&S MS not effective?
- What are the effective means of improving safety performance?
- If there is no evidence that OH&S MS are effective at improving safety, then shouldn't the resources that are usually directed to OH&S MS be diverted to more effective means of safety improvement?

A review of the literature[39] indicates that the failing of OH&S MS is neither due to the concept of systematic OH&S management nor the structure of the OH&S MS models used.

---

[38] Robson, L.S., Clarke, J.A., Cullen, K., Bielecky, A., Severin, C., Bigelow, P.L., Mahood, Q.,2007. *"The effectiveness of occupational health and safety management system interventions: a systematic review"*. Safety Science 45 (3), 329.

[39] Twelve (12) individual studies are cited in Podgorski. D., *"Measuring operational performance of OSH management system – A demonstration of AHP-based selection of leading key performance indicators"*, Safety Science 73 (2015) 146–166, p.146.

Their failure lies in the formal and frequently bureaucratic and paperwork-intensive approach of organisations, certification bodies and audit teams to ensure compliance of such OH&S MS with relevant criteria, without taking adequate consideration of the actual safety and health performance aspects of those systems.

Some studies indicate that low OH&S MS performance is related to insufficient knowledge and competence on the part of the auditors[40], and to the absence of verification and validation of auditor's competencies[41].

It is also claimed in studies that OH&S MS auditors focus on checking on the formal compliance of OH&S management procedures against pre-determined criteria, rather than delving to the core of the technical issues affecting safety - human factors, and the employer–employee relationship - which are the aspects that actually provide the foundation for actions for the benefit of safety[42].

Model Act or not, something needs to be done to rectify the failings of OH&S MS and the wasted resources often involved in operating them.

---

[40] Blewett, V., O. *"Weighing the pig never made it heavier: auditing OHS, social auditing as verification of process in Australia"*. Safety Science 49 (7), 1014.

[41] Dellacherie, C., 2010. *"La certification des enterprisedand le domaine de la sante autravail, Avis du Conseil economique, social et environmental. La Documentation francaise"*, Direction de lancaiseterprisedand le domaine de la s Republique Francaise. <http://www.ladocumentationfrancaise.fr/var/storage/rapports-publics/104000362/0000.pdf> (July 12, 2013).

[42] Fernandez-Muniz, B., Montes-Peon, J.M., Vazquez-Ordas, C.J., 2012. *"Occupational risk management under the OHSAS 18001 standard: analysis of perceptions and attitudes of certified firms"*. J. Clean. Prod. 24, 36.

It is not the scope of this book to suggest remedies to this situation, however it is suggested to PCBUs that in response to research on this subject, their OH&S MS should pay more attention to the three (3) factors just mentioned which have been shown to have a positive influence on safety performance, that is:

- Technical issues affecting safety (ie. the things that give rise to hazards / risks and the things that control hazards / risks).
- Human factors (ie. how and why humans behave physically and psychologically when at work and in the workplace).
- Employer–employee relationships (ie. how these multi-faceted relationships establish incentives or dis-incentives to safe behaviour and affect safety outcomes. This aspect also involves workplace culture and worker engagement in safety matters).

Safety Lifecycle Approach

The 'safety lifecycle' approach is a 'system safety'[43] or 'cradle-to-grave'[44] process whereby all of the risk controls that are necessary to keep a plant or system (eg. a car, a train, a mine, a factory, an aircraft etc…) safe are identified (ie. cradle), appropriately specified, designed, implemented, maintained and eventually decommissioned (ie. grave). These risk controls may typically be derived using the hierarchy of risk controls and include administrative controls such as rules and procedures, engineered controls such as electrical/electronic or

---

[43] System Safety is the application of scientific, technical and managerial skills to hazard identification, hazard analysis, and elimination, control, or management of hazards throughout the life-cycle of a system, program, project or an activity or a product.

[44] ie. from the "concept" phase of a system to "decommissioning / disposal".

mechanical control and protective systems, guards and barriers, as well as measures to substitute or eliminate risk.

AS(IEC)61508[45] is one of a number of industry standards[46] that applies a safety lifecycle process, as shown in Figure 4.3 (taken from Figure 1 of AS(IEC)61508.1).

---

[45] AS(IEC)61508 Functional Safety of Electrical / Electronic / Programmable Electronic Safety-related Systems).

[46] Others include CENELEC standard EN50126 (Specification and Demonstration of Railway Reliability, Availability, Maintainability and Safety) and AS(IEC)61511 (Functional Safety for the Process Industries).

Figure 4.3: AS61508.1 Safety Lifecycle

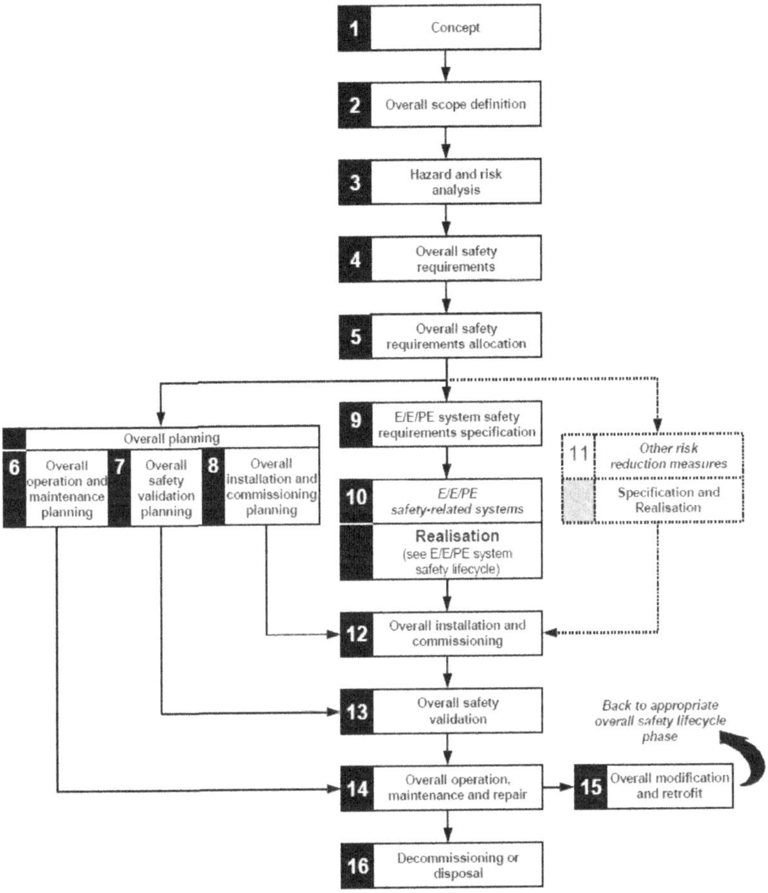

Each of the 16 lifecycle phases has its own detailed requirements, but the key features of the process are:

1. The process is initiated via planning of the safety lifecycle activities,
2. There is a hazard and risk analysis of the plant or system (known as the equipment under control, or EUC) and its control system in the specific application and environment,
3. Risk controls are identified during the hazard and risk analysis and their safety integrity (reliability) requirements are determined on the basis of risk,
4. Following this the designer / supplier of the risk controls is provided with a Safety Requirements Specification (SRS) detailing the functional requirements and safety integrity (reliability) requirements.
5. There is independent verification and validation during the design / installation / commissioning phases that the design of the risk controls meets the requirements of the Safety Requirements Specification.
6. A strong emphasis is placed on ensuring compliance to any maintenance requirements of the risk controls through-life,
7. A strong emphasis is placed on effective control of modifications through-life,
8. A strong emphasis is placed on the formal documentation of all activities,
9. A strong emphasis is placed on ensuring that those who perform safety lifecycle tasks are competent to do so.

Safety Case Guideline

In 2014 The Risk Engineering Society (RES) of Engineers Australia (EA) wrote, and published via Engineers Media Pty. Ltd., a practice note titled as follows:

"Safety Case Guideline, 3rd Edition – Clarifying the safety case concept to engineer due diligence under the provisions of the Model Workplace Health and Safety Act 2011"

The guideline claims to have been prepared by a committee of the RES, reviewed by a barrister and endorsed by the RES National Committee. The general thrust of the guideline and much of the text and graphical content appears to be common with material in a textbook written by two (2) members of the drafting committee[47].

In its summary (page 6) the guideline positions itself as a tool to *"positively demonstrate due diligence consistent with the requirements of the Model Work Health and Safety (WHS) legislation"*. The guideline claims to adopt a precautionary approach to work health and safety using *"the demonstration of due diligence as a defence against claims of negligence"* (also page 6). The guideline further claims (on page 6) that:

"Such a position should provide superior safety outcomes for all and offer protection against criminal charges for responsible officers under the provisions of the model WHS Act".

Further to this (on page 7) the guideline states that:

---

[47] Robinson, R., Francis, G., & Hurley, P., 2014, *Risk & Reliability – Engineering Due Diligence*, 9th Edition, R2A Due Diligence Engineers.

"The objective of this guideline is to provide practical guidance to engineers on preparing and presenting a safety case to positively demonstrate safety due diligence consistent with the requirements of the model Work Health and Safety Act".

In the following chapters the guideline details its 'due-diligence' approach to safety which, as it turns out, is a precautionary approach based on pre-emptively satisfying the principle of 'so far as is reasonably practicable' (SFAIRP) with respect to the implementation of risk controls (called 'precautions' in the guideline) to ensure safety. In doing so, this Safety Case Guideline provides a general framework for satisfying the primary duty of a PCBU, which is embodied by the SFAIRP principle.

However, the stated objective that it will allow a positive demonstration of 'due diligence', and the claim that it will offer protection to officers are not supported by its contents. As stated in a previous section, 'due diligence' is a duty to be exercised by the officers of a PCBU (Model Act, Section 27). This duty relates to strategic, structural, policy and resourcing decisions - that is, how the business or undertaking is run and how safety risk is managed in general. This is a fundamentally different duty to the primary duty of the PCBU itself (Model Act, Section 19), which goes to whether all 'reasonably practicable' measures have been taken to ensure the safety of workers and others against the specific hazards / risks posed by work and workplaces.

From the outset, it appears that the guideline's 'due diligence' approach is actually aligned with satisfying the primary duty of the PCBU, not the duty of the officers of the PCBU.

In detailing its approach, the guideline selects an alternative definition of 'due diligence' from a Comcare guideline entitled "Guidance for Officers in Exercising Due Diligence":

"Due diligence – in the context of work health and safety – means taking every precaution that is reasonable in the circumstances to protect the health, safety and welfare of all workers and others who could be put at risk from work carried out as part of the business or undertaking".

This definition of 'due diligence', at face value, appears to be more aligned with the Model Act definition of 'reasonably practicable' than the Model Act definition of 'due diligence'. The rest of this Comcare guideline goes on to detail how officers of a PCBU may successfully exercise their 'due diligence' duty. Rather than focussing on the above definition, it specifically addresses how officers can comply with each element of the Model Act definition of 'due diligence'.

The Comcare guideline definition of 'due diligence', when viewed in the context of the rest of the Comcare guideline and the intended definition of 'due diligence' of the Model Act, would appear to be an anomaly.

Clearly though, if the duty of the officer of a PCBU is to be satisfied, the definition of 'due diligence' in the Model Act should be the basis of the exercise of that duty.

The Safety Case Guideline therefore cannot, by definition, deliver on the claim that it will protect officers against prosecution under the Model Act, nor can it meet its objective - which is to provide for a positive demonstration of "due diligence" consistent with the Model Act.

Furthermore, with respect to the statement made on page 6 that the guideline uses *"the demonstration of due diligence as a defence against claims of negligence"* it is prudent that reference be made here to relevant case law.

In *Australian Iron and Steel Pty Ltd v Environment Protection Authority*[48], a decision of the Court of Criminal Appeal, the Court considered whether a defence of 'due diligence' was known to the Australian common law. The question arose in this case because such a defence is known to the Canadian common law and was relied upon by the appellant in the case. The Court of Criminal Appeal rejected the proposition and said that the defence of "due diligence" against claims of negligence is not recognised in Australian common law.

In summary of this section:

- In the Model Act, due diligence is a duty to be exercised by officers of a PCBU and is different to the primary duty of a PCBU.
- Satisfying the primary duty of a PCBU (ie. SFAIRP) does not automatically protect an officer of a PCBU from prosecution for failing to exercise their duty (ie. 'due diligence'), and vice-versa, because these are different duties.
- Whilst due diligence is a statutory duty under WH&S law, it is not available as a defence against claims of negligence in Australian common law.
- The *Safety Case Guideline* may assist in positively demonstrating that the primary duty of a PCBU (ie. SFAIRP) can be satisfied.
- The *Safety Case Guideline* <u>does not</u> positively demonstrate how the duty of an officer of a PCBU (ie. 'due diligence' as defined in the Model Act) can be satisfied.
- The *Safety Case Guideline* <u>cannot</u> assist in providing a defence against common law claims of negligence.

---

48 [1992] 29 NSWLR 497.

Control Effectiveness

Clearly, 'control effectiveness' – the technical capability of a risk control to effect the elimination or a net reduction in risk - goes to the 'suitability' of risk controls and so should be considered in order to satisfy the primary duty of a PCBU and ensure the most effective risk controls are implemented.

Control effectiveness evaluation may therefore assist in selecting the best option from a range of possible risk control options and may also assist in justifying why particular known and available risk controls should not be implemented. However, it must be remembered that "control effectiveness", of itself, cannot provide proof that the test of 'reasonably practicable' has been met.

Control effectiveness only answers the question as to how good a particular risk control or risk controls are against a hazard or risk. The test of 'reasonably practicable' requires an answer to a different question – have all the reasonably practicable risk controls been implemented?

Little is said about control effectiveness in the risk management standards apart from acknowledging its importance within the process of risk evaluation and treatment. The task of developing and publishing methodologies for systematically assessing control effectiveness has generally been left to others.

Two (2) informative references on control effectiveness evaluation are:

1. Australian Coal Association Research Programme (ACARP) Research Paper C15011, 2009[49], *"Acceptable Risk: A Matter of Establishing Adequate Control"*.
2. Australian Coal Association Research Programme (ACARP) Research Paper C23007, 2015[50], *"Selection and Optimisation of Risk Controls"*.

ACARP Paper C15011 details the Control Effectiveness Evaluation Method (CEEM), as shown in Figure 4.4.

Figure 4.4: ACARP CEEM Method

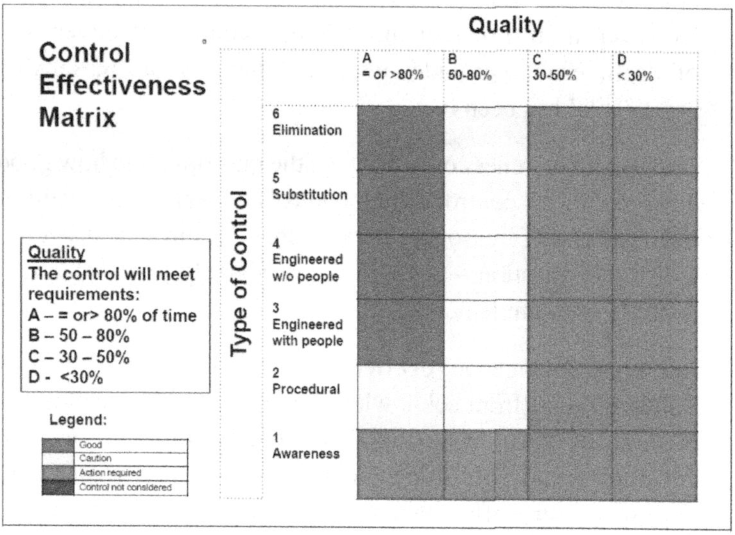

---

[49] Prof. Jim Joy and Dr. Guldidar V. Kizil, Minerals Industry Safety and Health Centre (MISHC), Sustainable Minerals Institute (SMI), University of Queensland.

[50] Dr. Maureen Hassall, Prof. Jim Joy, Dr. Chris Doran and Mr. Marcus Punch, Minerals Industry Safety and Health Centre (MISHC), Sustainable Minerals Institute (SMI), University of Queensland.

In Figure 4.4 control effectiveness is assessed via two (2) variables:

1. The quality of the risk control measured as a probability that it will meets its requirements when called upon.
2. The type of risk control, with respect to a hierarchy of risk controls, and

A more sophisticated method is proposed in ACARP Paper C23007, which is shown in Figures 4.5a and 4.5b. Control effectiveness is assessed via two (2) variables:

1. The quality of the risk control (Figure 4.5a), measured in terms of:
    a. The coverage / availability / reliability of the object / technological aspect of the risk control, and
    b. The coverage / availability / reliability of the human action aspect of the risk control, and
2. The impact of the risk control on risk (Figure 4.5b).

Both methods can evaluate the effectiveness of an individual risk control. They may therefore assist in selecting the most effective risk controls from a suite of risk control options. They may also assist in the justification of non-implementation of particular controls.

Figure 4.5a: Control Effectiveness Matrix I

CONTROL QUALITY		No object/ technology component to control	Coverage, availability and reliability of OBJECT/TECHNOLOGY component of control: Works in >95% of scenarios/areas	Works in 90%-95% of scenarios/areas	Works in 75%-90% of scenarios/areas	Works in 50%-75% of scenarios/areas	Works in <50% of scenarios/areas
No human action component to control			Excellent	Very Good	Good	Poor	Very poor
Coverage, availability and reliability of HUMAN ACTION component of control	Works in >95% of scenarios/areas	Excellent	Excellent	Very Good	Good	Poor	Very poor
	Works in 90%-95% of scenarios/areas	Very Good	Very Good	Good	Good	Poor	Very poor
	Works in 75%-90% of scenarios/areas	Good	Good	Good	Poor	Poor	Very poor
	Works in 50%-75% of scenarios/areas	Poor	Poor	Poor	Poor	Very poor	Very poor
	Works in <50% of scenarios/areas	Very poor	Very poor	Very poor	Very poor	Very poor	Very poor

Figure 4.5b: Control Effectiveness Matrix II

		CONTROL IMPACT - Degree to which the control impacts residual risks				
		Significant impact	Impact	Slight impact	No impact	Adverse impact
	CONTROL EFFECTIVENESS	Presence/action of control **significantly reduces** residual risk. Absence/failure of control **significantly increases** the residual risk	Presence/action of control **reduces** residual risk. Absence/failure of control **increases** the residual risk	Presence/action of control **slightly decreases** the residual risk. Absence/failure of control **slightly increases** the residual risk	Presence/action or absence/ failure of the control **does not change** the residual risk	Presence/action of control has potential to **increase** residual risk
CONTROL QUALITY from matrix one	Excellent	Highly effective	Very good effectiveness	Marginally effective	Poor effectiveness	Ineffective
	Very Good	Very good effectiveness	Very good effectiveness	Marginally effective	Poor effectiveness	Ineffective
	Good	Very good effectiveness	Very good effectiveness	Marginally effective	Poor effectiveness	Ineffective
	Poor	Marginally effective	Marginally effective	Poor effectiveness	Ineffective	Ineffective
	Very poor	Poor effectiveness	Poor effectiveness	Ineffective	Ineffective	Ineffective

During ACARP Project CC23007 a view was taken that to properly understand 'control effectiveness', an understanding was required of the fundamental properties of the 'risk controls' themselves.

Many definitions have been provided for 'risk control' and indeed many alternative terms are in use (eg. barrier, safeguard, precaution, control, treatment, layer of protection etc...). Fundamentally, the function of a risk control is to intervene in an accident sequence - to stop the accident sequence (ie. arrest it), or to deviate its propagation to a less severe consequence (ie. deflect it). This concept of a 'risk control' or 'barrier' is supported by the overall scheme of a 'bowtie' analysis as shown in Figure 4.6.

Figure 4.6: Bowtie Diagram

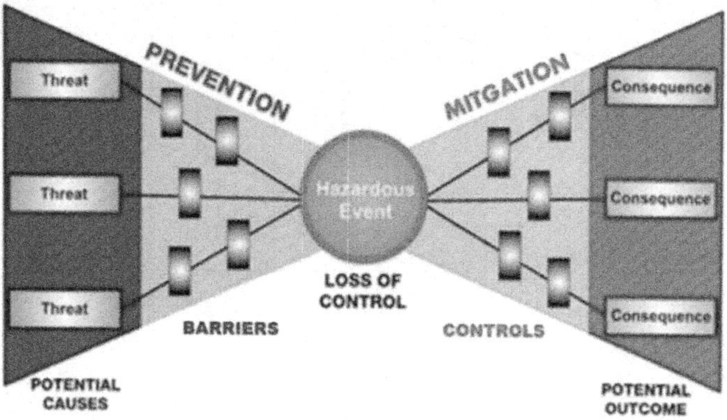

For example, the release of methane from coal mine workings poses a threat, which if not adequately controlled, can lead to a fire or explosion. This threat can be 'arrested' by preventive barriers / controls (eg. gas drainage from the coal seam prior to mining, ventilation of mine airways, etc...). These maintain control of the concentration of methane in air and ensure it remains outside of the flammable or explosive range. Should these measures fail to control methane concentrations, this would be a hazardous event. Further mitigating controls can be put in place to 'arrest' the accident sequence towards an explosion or 'deflect' the outcome to less severe safety consequences (eg. gas monitoring within the mine workings, explosion protected equipment, mine automation (no workers present), etc…).

In general, a risk control may be:

- A **tangible / physical object or system**, which of itself, arrests/deflects an unwanted event.

  It may be passive (eg. machine guarding) or active (eg. pressure relief valve).

  It may be automatically operated (eg. fire suppression system) or able to be manually operated (eg. manual plant shut-down).

  It may be preventive (eg. transformer fault protection) or mitigating (eg. transformer bund).

- A **human act** (eg. behaviour or response to stimuli), which of itself, arrests/deflects an unwanted event.

  It may be derived from the contents of a procedure, training or experience about what is expected of a person in a given situation.

  It may be preventive or mitigating.

  It can often be described using a verb / noun pair (eg. obey speed restrictions, isolate electrical supply, apply emergency brake, wear safety glasses, drink water).

The characteristics of an effective risk control are:

Pro-active – it prevents the unwanted event in the first instance, rather than controlling the consequences.

Potent (ie. efficacy) – it is technically capable of arresting or deflecting the accident sequence without imposing additional risk. (eg. guarding of the moving parts of machinery has greater efficacy than an emergency stop in the prevention of worker entanglement; morphine has greater efficacy than paracetamol in the treatment of pain relief, but poses an additional risk of addiction).

Responsive – it is in place prior to the occurrence it will be called upon to arrest or deflect, or operates within sufficient time (eg. good roadway design and construction is a control against car accidents that is put into place well before any car accident sequence occurs; the use of very early smoke detection alarms (VESDA) for early detection of fire in facilities).

Robust – it can cope with changes to the operating environment (eg. new people, new procedures or methods, changes to the physical environment).

Realistic – it represents value for money, is simple, and has ease of legacy (passing on of knowledge / experience).

Reliable – high probability of successful operation when required.

The reliability of a risk control is usually only relevant if the other criteria are met first. For instance, if the risk control is not technically capable of arresting or deflecting the accident sequence, or cannot do so in sufficient time, then no amount of reliability would be expected to improve its effectiveness.

For example, a machine emergency stop is generally a reactive risk control. No matter how reliable it is, it may be quite ineffective at prevention of entanglement of a worker in the moving parts of the machine. The effectiveness of the machine emergency stop against this hazard would not be improved by designing it to a higher level of reliability (eg. a safety integrity level, or SIL) because for this hazard it is neither a potent or responsive risk control. In this case, guarding of the moving parts of the machine would be a more effective risk control. Guarding is capable of arresting the accident sequence and it is responsive (it is put into place well before the hazardous event even arises).

However, if a risk control is known to be potent against a particular hazard and be deployed in sufficient time, then any improvement in the reliability of that risk control should improve the overall effectiveness of the control.

For example, a gas monitoring and shut-down system is known to be technically capable of promptly detect increased levels of methane in a mine and automatically stopping power to the affected areas. The effectiveness of the system against increased methane levels would generally be improved by designing it to a higher level of reliability because this improves the chance that the technical capabilities of the system are available when required.

A 'risk control' is often supported by things which help assure its effectiveness but sometimes these things are mistaken as being controls too. The key test of whether something is a risk control is whether it can stop the accident sequence (ie. arrest it), or to deviate its propagation to a less severe consequence (ie. deflect it). Examples of things that the author has observed being thought of as risk controls, but which are not, include:

- Training – the risk control is not the training but the human act/s or response/s that is the subject of the training.
- Procedures – the risk control is not the procedure but the human act/s or response/s that is required by the procedure.
- Competency assessment - the risk control is not the competency assessment but the human act/s or response/s that is being assessed.
- Maintenance tasks - the risk control is not the maintenance task but the function or feature of the tangible / physical object or system that is being maintained.
- Common-sense – this is a notion based on the vastly unfounded assumption that an arbitrary person can be expected to act or respond in a certain way in a particular set of circumstances without prior training or instruction as to what is required of them.
- Prayer (yes, I have seen this relied upon as a risk control) – to include prayer within the realm of legitimate risk controls would be to presume that the existence of an omniscient and omnipotent God is a matter of fact, rather than what even the religious know it to be - a matter of belief.

In practice, effective controls should be:

Evidence-based – the risk control is known (ie. by history, research, testing etc…) to be effective against the hazard or cause to which it is being applied as a defence.

Specifiable – the parameters or underlying requirements which make the risk control effective against the hazard or cause to which it is being applied can be described in a way that allows others to replicate the risk control.

Measureable – the parameters or underlying requirements which make the risk control effective against the hazard or cause to which it is being applied can be measured and compared to what was specified.

Auditable – the sustained and correct application of the risk control and the resultant effect on safety can be monitored, for the purposes of review and improvement.

Figure 4.7 illustrates these principles. Evidence is at hand that if a person is struck by a motor vehicle the chance of them being killed rises significantly if the speed of the motor vehicle is above 40 km/hr. There is an increased risk of people being struck by motor vehicles near schools in the morning and afternoon when people and motor vehicles converge near schools to drop and pick-up children from school.

There are a number of risk controls that can be applied to this risk. In the event of a collision with a person though, the slower the motor vehicle is travelling, the better the outcome for the injured person. In this case a time-based speed limit is imposed near schools during morning drop-off and afternoon pick-up times. The speed restrictions may be communicated to motorists in a number of ways - signage, flashing lights, speed feedback displays, community awareness campaigns in the media etc…

Compliance to this risk control can be measured. An effective way of doing this is to place speed measurement devices and cameras in the areas where the speed limit applies. Following a period of operation, the effectiveness of the control can be reviewed. Road traffic authorities may review all accidents in school zones involving a person being struck by a motor vehicle. The effectiveness of the risk control might be gauged by calculating the proportion of school-zone accidents which result in fatalities.

Figure 4.7: Effective Controls

Evidence-based, Specifiable, Measurable and Auditable

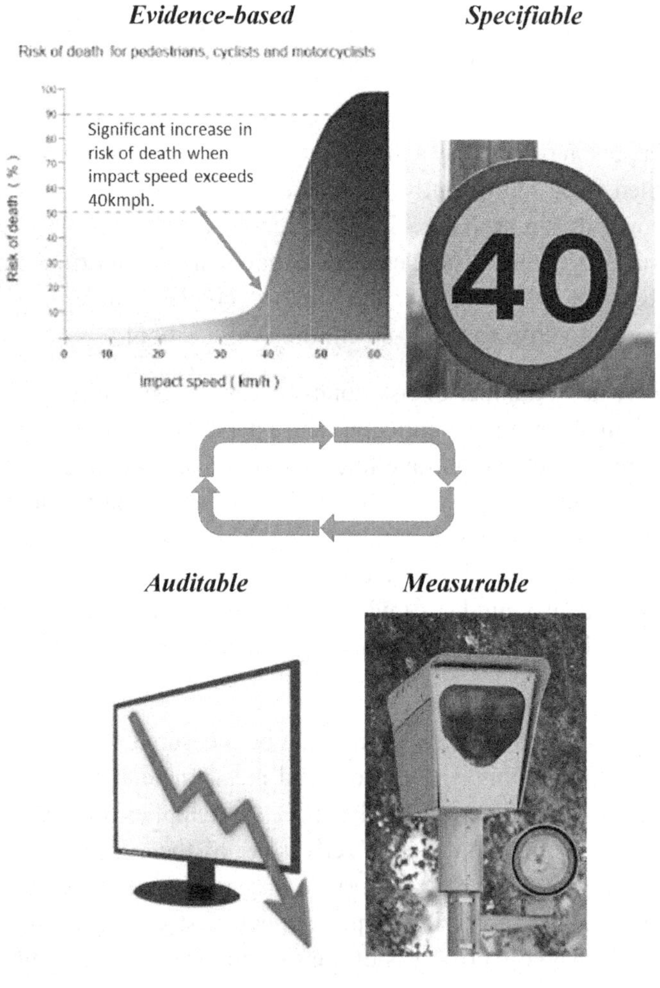

Chapter 5 - A Suggested Framework

Concept for Safety Risk Management

This chapter briefly outlines a safety risk management approach that applies the legal and risk management concepts and processes discussed in earlier chapters of this book.

The concept for safety risk management with respect to the Model Act is shown in Figure 5.1.

Figure 5.1: Safety Risk Management Concept

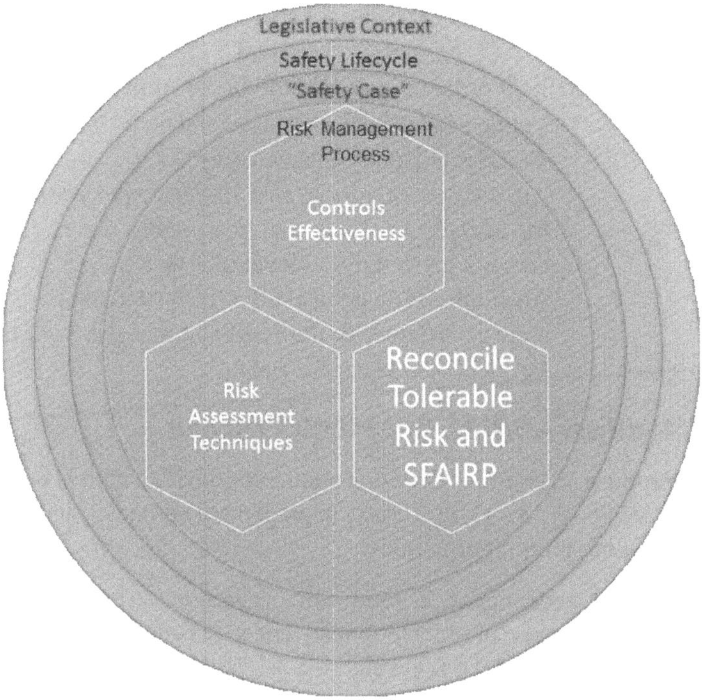

This concept is based upon the following key principles:

- Acknowledgement of the legislative / regulatory context of the Model WH&S Act.
- Taking a lifecycle-management approach to safety (ie. 'cradle to grave' or system safety approach).
- The use of 'safety case' concepts, particularly the pre-emptive justification of the use (and non-use) of particular risk controls and the construction of arguments that an acceptable level of safety is achieved.
- The use of a risk management process derived from the generic AS/NZS ISO31000 risk management process, tailored to meet the requirements of the Model Act and its legislative / regulatory context. Within this risk management process, there should be an emphasis on:
    - Using appropriate risk assessment techniques to identify and analyse hazards / risks.
    - An acknowledgement that the acceptance criteria for hazards / risks is the "reasonably practicable" test, not a pre-defined level of risk.
    - An emphasis on control effectiveness evaluation as a means to identify the most suitable risk controls for identified hazards / risks.

Modified AS/NZS/ISO31000 Process

I propose an overall safety management process which is based on AS/NZS ISO31000 risk management process and the following guiding principles:

- The acceptance criteria for hazards / risks is the "reasonably practicable" test, not a pre-defined tolerable (or acceptable) risk level.
- Hazard / risk identification must systematically identify all 'reasonably foreseeable' hazards / risks that may harm people.

- Risk analysis must involve a determination of the causes, likelihood of occurrence, and potential consequences of the hazards / risks.
- Risk evaluation and treatment should follow the modified process described in Chapter 4, which is shown again on the next page.

Modified AS/NZS ISO31000 Risk Evaluation / Treatment Steps

1. Determine all known and available risk controls for each hazard / risk.
2. Decide which risk controls are and are not "reasonably practicable" with particular attention to the following:

   a. The suitability of each risk control for the circumstances – taking account of their effectiveness against the risk at hand (see the next section), further risks introduced etc…

   b. Risk elimination measures must be used against a hazard, where "reasonably practicable", before risk reduction measures are considered.

   c. If risk elimination measures are not "reasonably practicable", the hierarchy of risk controls must be used to guide the selection of the most effective risk reduction measures for each hazard.

   d. Justification should be provided for each risk control determined to be not "reasonably practicable" to implement – eg. ineffective in the circumstances, gross disproportionality, net risk increase.

3. Decide which "reasonably practicable" risk controls are to be implemented and not implemented, noting that if a PCBU does not have the capacity to pay for a reasonably practicable risk control, it should cease or not undertake the activity which gives rise to the risk.

4. Determine the reliability, integrity, strength or robustness requirements of any risk controls which require a tolerable risk approach to their specification and design (eg. earthing, electronic safety systems, explosion protection, and structural design). This should also involve a consideration of the level of reliability / integrity / strength / robustness that is reasonably practicable for a designer to achieve, and take account of the specific requirements of regulations or standards for such risk controls.

The following principles (already mentioned in Chapter 3) should be observed when decisions are made regarding the acceptability / tolerability of hazards / risks:

- The level of risk which remains when all of the selected risk controls for a particular hazard are put into place cannot be used to justify a position that nothing else needs to be done, or that there are no further 'reasonably practicable' risk controls, nor can it be used to justify the acceptance of the risk.
- In particular, for low level risks, the level of risk which would exist without having any risk controls in place for a particular hazard cannot be used to justify a position that it is not necessary, or reasonably practicable for any risk control to be implemented for a particular hazard.
- It may be difficult for a PCBU to justify not implementing a risk control on the basis of cost. If the degree of harm is significant (ie. serious injury or death) a "gross disproportionality" argument discounting the use of a known, available and suitable risk control is unlikely to be justifiable[51].
- Aspects of a PCBUs financial position (particularly their incapacity to pay for a risk control) cannot be taken into account in determining what is reasonably practicable. In order to comply this could mean that the activity giving rise to the hazard or risk would need to cease or not be undertaken.

Particular attention should be paid to the effectiveness of controls as this will form a large part of the justification of the

---

[51] See Johnstone, R. and Tooma, M., 2012, *"Work Health and Safety Regulation in Australia – The Model Act"*, The Federation Press, p.74.

selection of risk controls that are implemented and those that are not.

Furthermore, when a specialist approach involving 'tolerable risk' level targets is to be used, care must be taken to ensure all reasonably practicable risk controls are selected before this approach is undertaken. The last section of this chapter provides more details about how to proceed with a tolerable risk approach.

Such a modified risk management process is shown in Figure 5.2.

Figure 5.2: Modified Risk Management Process

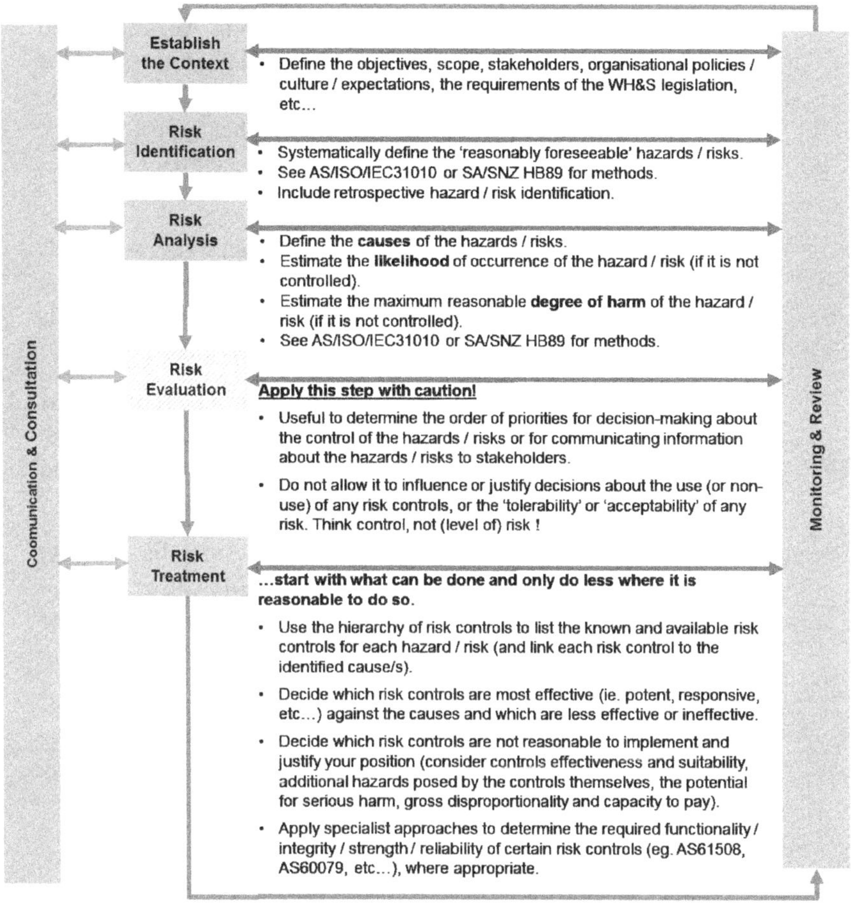

Using 'Tolerable Risk' Standards

As previously mentioned, a number of current Australian and International standards, codes of practice and guidelines use the notion of tolerable risk as a basis for determining the reliability, integrity, strength, robustness or other performance requirements of risk controls for particular hazards. Examples of this are:

- AS(IEC)61508[52] and associated functional safety standards use a notion of tolerable risk level to determine the reliability (safety integrity) requirements of risk controls based on electrical and electronic technologies.
- AS(IEC)60079 series of standards for hazardous areas which determine the requirements for explosion protection measures for a hazardous area on the basis of the likelihood of a flammable gas or dust becoming present in that area.
- Energy Networks Association of Australia (ENA) Guideline EG-0 power system earthing guide uses a notion of tolerable risk level to determine the requirements for electrical earthing of installations.

Nevertheless, in those instances where a 'tolerable risk' approach is to be used in a jurisdiction that is subject to the Model Act (whether mandatorily or voluntarily), it is recommended that the following caveats are applied:

- PCBUs must obtain a level of confidence that their primary duty is met by the selected regime of risk controls for a hazard before applying a 'tolerable risk' approach to determine the reliability, integrity, strength, robustness or

[52] Some Australian standards are referred to throughout this book as "AS(IEC)…". This was done for the sole purpose of expressing their origin and equivalence to the IEC standards.

other performance requirements of particular risk controls within that regime.

- Tolerable risk criteria should be selected that are reasonable and commensurate with the potential severity of the consequences of the hazards / risks being considered.
- The determination of performance requirements on the reliability, integrity, strength or robustness or other performance requirements should also take account of what would be "reasonably practicable" for a designer or supplier to achieve.
- Where a regulation, code of practice or standard specifies a level of reliability, integrity, strength or robustness or other performance requirements for a particular type of risk control, and this is more onerous than the performance requirements derived via a 'tolerable risk' approach, then the requirement of the regulation, code or standard should take precedence.

Each of the above-mentioned standards and other documents detail the techniques to be used in applying a 'tolerable risk' approach. It is not the scope of this book to duplicate the details of these techniques here. However, one question remains unanswered by the preceding discussion, which is – what would a reasonable "tolerable risk" target be?

The following information is provided as guidance in those circumstances where a "tolerable risk" approach is recommended or mandated. The use of a "tolerable risk" approach must not supplant the requirements of the PCBU's primary duty (ie. SFAIRP).

"Tolerable risk" targets have already been suggested by various organisations and industry groups throughout the world. The United Kingdom Health & Safety Executive (UK H&SE) and the United Kingdom Institution of Gas Engineers and Managers (IGEM) are two groups who have been at the forefront of these developments. These two bodies have suggested that the Individual Risk (of death) Per Annum (IRPA) to a worker should be no more than 1 in 1,000 (or 10^{-3}) considering all hazards that they are potentially exposed to whilst at work. They have also suggested that an IRPA of less than 1 in 1,000,000 (or 10^{-6}) could be considered to be "negligible" or "broadly acceptable"[53].

In some cases the IRPA targets have been extended to a broader definition of "risk per annum" targets for a range of consequence severities. For instance, in EN50126[54] safety targets are expressed for railway developments by way of an 'F-N curve', as shown in Figure 5.3. An F-N curve is a graphical representation of the probability of a specified consequence occurring within a specified population[55].

---

[53] See Institution of Gas Engineers and Managers, 2010, *IGEM SR/15 - Integrity of Safety-related Systems in the Gas Industry*, 5th Edition p.5 and UK Health and Safety Executive, 2001, *"Reducing Risk, Protecting People - HSE's decision-making process", HSE Books* pp.44-49.

[54] CENELEC standard *EN50126 - Specification and Demonstration of Railway Reliability, Availability, Maintainability and Safety (RAMS).*

[55] See *SA/SNZ HB89 Risk Management – Guidelines on Risk Assessment Techniques* or *IEC31010 Risk Management – Risk Assessment Techniques* for further details.

Figure 5.3: Tolerable Risk Graph

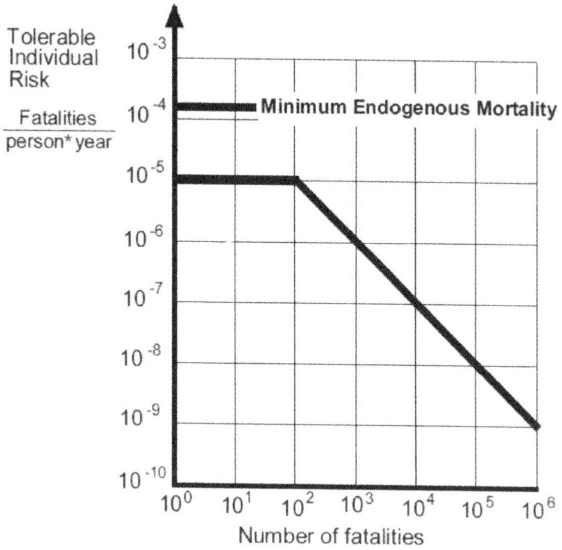

The term 'Minimum Endogenous Mortality' (MEM) used in this diagram relates to the underlying individual risk of death to certain age groups within a population as a result of day-to-day activities such as employment, sport, home activities and transport. In developed countries it is assumed that the age group with the minimum individual risk during these activities is the 5-15 year age group. Their individual risk per annum, or IRPA, for these activities has been determined to be 1 in 5,000, or 2.0×10^{-4}. EN50126 uses the MEM concept to set the overall risk targets for a new railway system. The risk posed by a new railway system should not significantly augment the underlying MEM in a population arising from its day-to-day activities. A 5% increase in mortality is considered to be significant.

Therefore, the single fatality IRPA target for the railway system is set at 5% of MEM, which is 5% of 2.0×10^{-4}, or 1.0×10^{-5}.

Figure 5.3 also accounts for a phenomena in societal risk[56] known as 'differential risk aversion' whereby the tolerable frequency of an adverse event decreases as the severity of the consequences of the event increases. Differential risk aversion accounts for the fact that high consequence events are less socially and politically acceptable than low consequence events.

In Australia, tolerable risk targets have also been suggested, including:

- NSW Department of Planning *"Risk Criteria for Land Use Planning"* (2008) suggests risk acceptability criteria for new installations within certain land-use zones on the basis of IRPA; as follows:

  5.0×10^{-5} in areas zoned as industrial areas.

  1.0×10^{-5} in areas zoned as sporting complexes and outdoors active areas.

  5.0×10^{-6} in areas zoned as commercial areas (eg. shopping centres).

  1.0×10^{-6} in areas zones for residential, hotels / motels and tourist facilities.

  5.0×10^{-7} in areas zoned as hospitals, schools, child and aged care facilities.

- Minerals Industry Safety and Health Centre (MISHC) *"Final Report to Queensland Resources Council (QRC) on*

---

[56] Societal risk aims to provide a single measure of the chance of accidents that could harm a number of people at the same time, such as during catastrophic industrial accidents.

underlying Causes of Fatalities and Significant Injuries in the Australian Mining Industry", March 2005:

"Safe" industry should be about 0.005 fatalities / million man-hours, or a factor of ten lower than currently reported in the mining industry.[57]

Given a fatality rate of 0.005 fatalities per million man-hours, and assuming the average individual mine worker performs 2,000 hours of work per year, this figure translates to an IRPA of 10^{-5}.

The concept of these targets is shown in Figure 5.4.

Figure 5.4: Individual Risk Targets

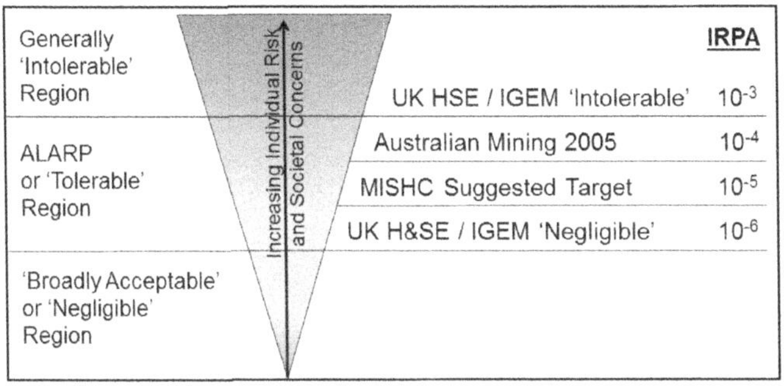

Regarding recent work safety performance in Australia, Table 12 of Safe Work Australia's report *"Work-related Traumatic Injury Fatalities, Australia 2013"* (July 2014), shown below in Table 5.1, details the fatality rates of Australian workers between 2003 and 2013.

---

[57] Minerals Industry Safety and Health Centre (MISHC), 2005, *"Final Report to Queensland Resources Council (QRC) on underlying Causes of Fatalities and Significant Injuries in the Australian Mining Industry"*, p.11.

Table 5.1: Australian Industry Fatality Rates

Table 12: Worker fatalities: fatality rate (fatalities per 100 000 workers) by industry of employer, 2003 to 2013 (sorted by 2013)

Industry of employer	2003	2004	2005	2006	2007	2008	2009	2010	2011	2012	2013	Total
Agriculture, forestry & fishing	16.39	21.56	17.45	13.11	14.74	18.97	17.01	14.82	18.05	15.91	15.11	16.66
Agriculture	*14.51*	*17.34*	*14.38*	*11.42*	*13.00*	*14.19*	*13.89*	14.04	*15.76*	11.99	13.27	13.99
Transport, postal & warehousing	14.43	13.56	10.99	12.21	15.32	12.41	10.85	7.63	8.23	10.65	7.76	11.14
Road Transport	*19.89*	*21.54*	*18.61*	*22.35*	*25.34*	*21.35*	*22.30*	*14.08*	*12.15*	*21.58*	15.99	19.60
Road Freight Transport	*25.89*	*31.52*	*26.98*	*31.52*	*37.56*	*28.84*	*29.44*	*19.15*	*15.29*	*27.80*	*20.46*	*26.66*
Electricity, gas, water & waste services[b]	5.43	8.72	5.96	7.48	3.77	3.22	3.71	0.70	1.32	1.97	3.35	3.77
Arts & recreation services	2.05	3.20	3.46	2.82	5.13	1.58	0.49	2.60	2.84	2.36	3.34	2.71
Mining[a]	12.35	10.82	7.51	11.32	5.07	7.25	6.10	2.66	2.70	3.01	2.98	5.50
Construction	5.71	4.38	3.51	4.70	4.84	3.81	3.61	4.03	3.97	3.00	1.85	3.89
Rental, hiring & real estate services[b]	2.36	1.13	3.34	2.04	4.97	1.95	1.08	2.06	1.01	1.42	1.52	2.08
Other services[b]	0.94	1.67	2.45	1.67	0.89	1.52	0.90	0.43	0.89	0.65	1.50	1.21
Wholesale trade	1.85	2.61	2.67	2.02	2.04	2.70	0.97	2.40	0.72	2.19	1.40	1.95
Public administration & safety	1.42	1.70	2.11	1.92	1.87	0.57	1.51	0.80	1.41	1.05	1.36	1.53
Administrative & support services	1.14	2.63	3.37	2.26	3.35	2.90	3.35	3.09	1.99	1.50	1.27	2.42
Manufacturing	1.92	2.19	2.23	2.84	2.40	2.83	2.27	2.11	2.28	1.87	1.07	2.19
Retail trade	0.62	0.45	0.59	0.85	1.40	0.49	0.41	0.58	0.16	0.41	0.81	0.62
Accommodation & food services[b]	0.77	0.60	0.00	1.20	0.85	0.28	0.41	0.27	0.13	0.13	0.25	0.43
Health care & social assistance[b]	0.33	0.31	0.30	0.19	0.00	0.18	0.50	0.32	0.15	0.15	0.21	0.24
Professional, scientific & technical services[b]	0.32	0.95	0.29	0.81	0.93	0.63	0.50	0.47	0.23	0.88	0.11	0.55
Information media & telecommunications[b]	0.45	0.45	0.00	0.41	1.23	0.00	0.45	1.40	1.43	0.88	0.00	0.61
Education & training[b]	0.28	0.14	0.28	1.07	0.52	0.25	0.24	0.23	0.23	0.45	0.00	0.33
Financial & insurance services[b]	0.29	0.00	0.00	0.26	0.00	0.74	0.75	0.00	0.00	0.00	0.00	0.18
All industries	**2.72**	**2.94**	**2.57**	**2.77**	**2.93**	**2.59**	**2.35**	**2.03**	**1.96**	**1.98**	**1.64**	**2.39**

[a] Mining fatalities include those that occur in coal and metal ore mining, oil and gas extraction, sand and gravel quarrying, exploration and support services.
[b] Movements in fatality rates in industries with 5 or fewer fatalities each year should be viewed with caution.

For example, the fatality rate for mine workers has averaged 5.5 fatalities per 100,000 workers over the period, but this occurred within a set of annual results that varied significantly (ie. 2.7 to 12.4 fatalities per 100,000 workers). The average fatality rate for the period 2003-2013 translates to an IRPA of 5.5×10^{-5}. Some gains have been made since 2005, but the IRPA of 10^{-5} derived from MISHC's suggested target has not yet been met.

The MISHC "safe" industry target therefore still represents a substantial improvement on current safety performance (ie. 5.5 times better) in the mining industry and recent gains indicate that it is an achievable target in the medium-term.

In contrast to this is the road freight transport industry. The fatality rate for road freight transport workers has averaged 26.7 fatalities per 100,000 workers over the period, but this also occurred within a set of annual results that varied significantly (ie. 15.3 to 37.6 fatalities per 100,000 workers). The average fatality rate for the period 2003-2013 translates to an IRPA of 2.7×10^{-4}.

On the basis of these figures, workers in the road freight transport industry have a fatality rate that is 4.8 times greater than workers than the mining industry!

These risk targets consider the overall risk to a worker considering all hazards. However in practice, risks are assessed on an individual hazard / risk basis and only consider those workers who are potentially exposed. For the purposes of the 'tolerable risk' approach these IRPA targets need to be translated from a 'per annum' basis to a 'per hazard' basis.

A scheme for doing this is provided in Table 5.2. This table allows the derivation of a "per hazard" safety target, taking account of the chosen IRPA target (in this case 10^{-5} per year), the relative size of the workforce and the relative number of hazards posed to the workers.

Table 5.2: "Per Hazard" Safety Targets

IRPA Target (fatalities / employee / year)	No. of Workers	Overall Fatality Target (fatalities / year)	No. of Fatal Hazards Present	Per Hazard Frequency Target (events / year)
10^{-5}	1	10^{-5}	1	10^{-5}
			10	10^{-6}
			100	10^{-7}
			1000	10^{-8}
	10	10^{-4}	1	10^{-4}
			10	10^{-5}
			100	10^{-6}
			1000	10^{-7}
	100	**10^{-3}**	10	10^{-4}
			100	**10^{-5}**
			1000	**10^{-6}**
	1000	**10^{-2}**	10	10^{-3}
			100	**10^{-4}**
			1000	**10^{-5}**
	10000	10^{-1}	10	10^{-2}
			100	10^{-3}
			1000	10^{-4}

For example, for a typical large work site (such as a mine or other heavy industrial facility) with 100 to 1000 workers continuously present (bolded), and with 100 to 1000 fatal hazards (bolded) posed to each worker the suggested "per hazard" frequency target is 10^{-4} to 10^{-6} per year (bolded). A nominal target of 10^{-5} per year (1 in 100,000 per year) is suggested.

Furthermore, for a typical small work site (such as a retail shop) with 1 to 10 workers continuously present and with 1 to 10 fatal hazards posed to each worker, the suggested "per hazard" frequency target is 10^{-4} to 10^{-6} per year. A nominal target of 10^{-5} per year (1 in 100,000 per year) is suggested.

The "per hazard" target of 10^{-5} per year is anecdotally supported by mining industry guidance provided in MDG1010 (1997). An extrapolation of the safety risk matrix suggested in MDG1010 (1997), Figure A.9.2 (shown here in Table 5.3) indicates that for a single fatality consequence, a "Low" risk would only be achieved with a hazardous event frequency less than once per 100,000 years, (or, 10^{-5} per year).

The 10^{-5} per year target might be adjusted for those hazards which could result in multiple fatalities, or in less severe safety consequences such as minor injuries. This allows for differential risk aversion to be taken into account in the tolerable risk targets.

For example, a hazard with the potential for one (1) medically treatable injury (MTI) could be assigned the "per hazard" frequency target of 10^{-1} per year (1 in 10 per year), a hazard with the potential for ten (10) fatalities could be assigned a target of 10^{-6} per year (1 in 1,000,000 per year) and a hazard with the potential for one hundred (100) fatalities could be assigned a target of 10^{-7} per year (1 in 10,000,000 per year).

Table 5.3: Semi-quantitative Risk Matrix - Mining

	1 x Medically Treatable Injury (MTI)	1 x Compensible Injury (CI) or 10 x MTI's	10 x CI's	1 x Permanent Disablement (PD)	1 x Fatality	10 x Fatalities
Frequent ≥ 1 per year	Medium	High	Very High	Severe	Severe	Severe
Possible: < 1 per year (but ≥ 0.1 per year)	Medium / Low	Medium	High	Very High	Severe	Severe
Unlikely: < 0.1 per year (but ≥ 0.01 per year)	Low	Medium / Low	Medium	High	Very High	Severe
Very Unlikely: < 0.01 per year (but ≥ 0.001 per year)		Low	Medium / Low	Medium	High	Very High
Barely Credible: < 0.001 per year (but ≥ 0.0001 per year)			Low	Medium / Low	Medium	High
$< 10^{-4}$ per year				Low	Medium / Low	Medium
$< 10^{-5}$ per year					Low	Medium / Low
$< 10^{-6}$ per year						Low

Further anecdotal support is provided by the Hong Kong MTRC (rail authority) risk matrix shown in Table 5.4, which provides similar tolerable risk frequencies for a range of consequence severities.

Table 5.4: Semi-quantitative Risk Matrix - Rail

			CONSEQUENCE						
			7	6	5	4	3	2	1
			Trivial	Negligible	Marginal	Serious	Critical	Catastrophic	Disastrous
Staff/Contractor Safety	Fatality						<5	5 or more	
	Major Injury					<5	5 or more		
	Minor Injury	with ≥ 3 days sick leave			<5	5 or more			
		with < 3 days sick leave		<5	5 or more				
Passenger/Public Safety	Fatality						<5	5-50	51-500
	Major Injury					<5	5-50	51-500	501 - 5000
	Minor Injury				<5	5-50	51-500	501 - 5000	>5000
Service	System Disruption				<20 min	1 hour	1 day	· 1 week	1 month
	Line Disruption			20-60min	few hours	1 day	1 week	1 month	few months
	Station Disruption		<20min	few hours	1 day	1 week	1 month	few months	1 year
FREQUENCY	A Few times per week or more	≥ 100 /year	R3	**R1**	**R1**	**R1**	**R1**	**R1**	**R1**
	B Few times per month	≥ 10 - <100 /year	R4	**R1**	**R1**	**R1**	**R1**	**R1**	**R1**
	C Few times per year	≥ 1 - <10 /year	R4	**R2**	**R1**	**R1**	**R1**	**R1**	**R1**
	D Few times in 10 years	≥ 0.1 - <1 /year	R4	R3	**R2**	**R1**	**R1**	**R1**	**R1**
	E Once since operation	≥ 1E-2 - <1E-1 /year	R4	R3	R3	**R2**	**R1**	**R1**	**R1**
	F Unlikely to occur	≥ 1E-3 - <1E-2 /year	R4	R4	R3	R3	**R2**	**R1**	**R1**
	G Very unlikely to occur	≥ 1E-4 - <1E-3 /year	R4	R4	R4	R3	R3	**R2**	**R1**
	H Remote	≥ 1E-5 - <1E-4 /year	R4	R4	R4	R4	R3	R3	**R2**
	I Improbable	≥ 1E-6 - <1E-5 /year	R4	R4	R4	R4	R4	R3	R3
	J Incredible	< 1E-6 /year	R4	R4	R4	R4	R4	R4	R3

Chapter 6 - A Final Word...

The purpose of this book, as stated in Chapter 1, is to:

- Explain the essential concepts, principles and requirements for ensuring health and safety that are contained in the Model Act.
- Relate them to the established frameworks for health and safety management and risk management.
- Expose authoritative sources (and unauthoritative ones) for interpretation and learned opinions on the subject matter.
- Provide a critique on a number of contemporary health and safety management approaches through the prism of the Model Act's principles, concepts and requirements.
- Propose an enhanced framework for health and safety-related risk management,
- and in doing so, provide the means by which readers can critically evaluate information that they might receive on the subject.

This book has laid much focus on the "reasonably practicable" test which lies at the heart of a PCBU's primary duty in WH&S law. It was written in the knowledge that the Model Act is firmly in place and that "reasonably practicable" is a concept with significant history and is well ensconced within WH&S law in many jurisdictions.

The book was written to help those to whom the requirements of the Model Act apply. It was not intended to critique or evaluate the overall "reasonably practicable" philosophy and approach to safety, nor would it serve much purpose to the reader if it did.

However, the book would not be complete without at least asking (and attempting to answer) the following questions:

- ***Under what circumstances might a PCBU have identified all reasonably foreseeable hazards / risks and met the "reasonably practicable" test for all of them but the work or the workplace remains unsafe?***

- ***Under what circumstances might the Model Act's focus on risk control stifle societies' "appetite" for opportunity and risk-taking?***[58]

To answer the first question it must be understood that "reasonably practicable" is essentially an "equity" (ie. fairness) and "utility" (ie. usefulness) based criteria for making decisions about safety-related risks.

Reasonably practicable promotes "equity" because it supports the premise that all individuals have unconditional rights to certain levels of protection. Worker safety must be ensured so far as is reasonably practicable. The primary duty applies equally to all PCBUs and no PCBU owes a lesser duty than any other.

Reasonably practicable also promotes "utility" because it implies an approach to safety management that strives to find the most suitable and effective risk controls, and allows for certain risk controls to be discounted on the basis of non-effectiveness, unsuitability or gross disproportionality of cost to risk benefit.

However, the "reasonably practicable" test can only be applied to hazards / risks which are known.

---

[58] "Risk appetite" is the level of risk that an organization is prepared to accept before action is deemed necessary to reduce it. It represents a balance between the potential benefits of innovation and the threats that change inevitably brings.

A fundamental characteristic of a risk is that it can be named and described. The ability to control a risk only arises when it is identified. Whether an identified risk will actually occur is subject to uncertainty. Nevertheless, it is considered to be a risk because it is possible. For example, a motorist has the potential to have a motor vehicle accident on their way home from work. This event may or may not ever happen, but it could, and if it does happen, there is considerable uncertainty as to when it will happen and the degree of harm caused to the motorist and other road-users involved. Risks are “known unknowns”. “Known unknowns” can be named and described and from this arises an ability to proactively control them.

If a risk is not known, then it cannot be controlled. “Unknown unknowns” are not risks because they have not yet been discovered and therefore cannot be named or described. They are beyond the ‘state of the art’ of our knowledge and experience. “Unknown unknowns” therefore cannot be proactively controlled.

Therefore, it is theoretically possible that if all the “reasonably practicable” risk controls for the “reasonably foreseeable” “known unknowns” of a workplace are in place, there may still be some “unknown unknowns” that are yet to be discovered which will actually render the workplace unsafe until they are discovered and subject to control.

This question may appear hypothetical or philosophical at the outset, but consider the circumstances when new technologies are introduced to workplaces using an historical example.

Asbestos has been used by humans for at least 4,500 years[59]. Despite suspicions dating back to Roman times, a causal link

---

[59] Evidence shows that inhabitants of the Lake Juojärvi region in East Finland strengthened earthenware pots and cooking utensils with the asbestos mineral anthophyllite. See Ross, Malcolm & Nolan, Robert P (2003). "*History of asbestos discovery and use and asbestos-related*

between this mineral and certain respiratory diseases was not established until the late nineteenth century or early twentieth century[60]. In fact, the disease "asbestosis" was not clinically named until 1927[61].

This means that, until some point in the late 1800s asbestos risks were virtually "unknown unknowns" – they could not be named or described - so nothing was able to be done about them. So until the 1900s, the known and available risk controls for asbestos did not include many of the measures that we now know are necessary to adequately protect workers and others from asbestos-related diseases.

However as the 'state of the art' knowledge regarding asbestos evolved in the twentieth century, these "unknown unknowns" became "known unknowns" and with this arose the ability to control the previously unknown risks.

The central issue is that even if all of the known and available risk controls for asbestos of the 1800s were put in place at a

---

disease in context with the occurrence of asbestos within the ophiolite complexes". In Dilek, Yildirim, Newcomb, Sally. *Ophiolite Concept and the Evolution of Geological Thought*. Special Paper 373. Boulder, Colorado: Geological Society of America.

[60]Deane, Lucy (1899). *"Report on the health of workers in asbestos and other dusty trades" in HM Chief Inspector of Factories and Workshops, 1899, Annual Report for 1898*. HMSO London. pp. 171–172.

D. Auribault, *"Note sur l'Hygiène et la Sécurité des Ouvriers dans les Filatures et Tissages d'Amianté (On hygiene and security of the workers in the spinning and weaving of asbestos)"* in Le Bulletin de l'Inspection du Travail, 1906, pp. 120–132.

H. M. Murray, testimony before the Departmental Committee on Compensation for Industrial Diseases *"Minutes of Evidence, Appendices and Index"*, 1907. p. 127.

[61] Cooke, W. E. (1927). *"Pulmonary Asbestosis"*. *BMJ* **2** (3491): 1024–5.

workplace during the 1800s, asbestos production and use remained unsafe. A significant "unknown unknown" was not being controlled. One of the worst aspects of the history of asbestos is that even after the risks became well known, the effort to control them was unnecessarily delayed until well into the latter half of the twentieth century (see Figure 6.1).

Figure 6.1: Risks are "Known Unknowns"
Risk controls cannot address the "unknown unknowns". They must become "known unknowns" in order that they can be controlled.

Source: Popular Science, January 1956.

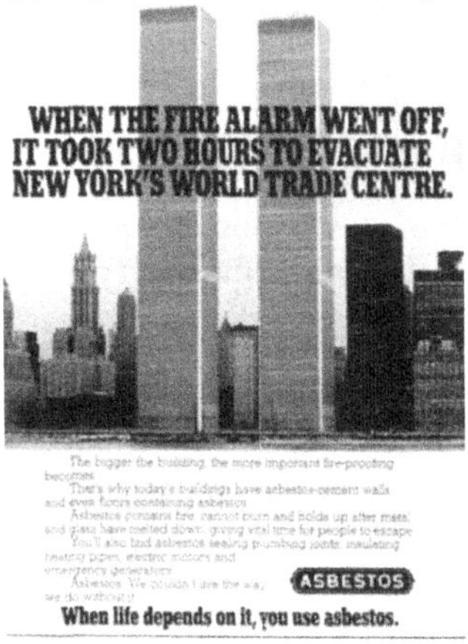

Source: Asbestos Magazine, November 1981.

This phenomena might also apply whenever new technologies or work methods are introduced to workplaces. Even if the risk controls for the known hazards / risks comprise everything that is "reasonably practicable", this says nothing of the risk posed by any yet unknown hazards.

The "reasonably practicable" test cannot ensure safety against the unknown, which may be eventually found to be significant enough that the work or workplace were actually unsafe.

To answer the second question…

'Tolerable risk' (and acceptable risk) refers to a willingness by society as a whole to live with a risk so as to secure certain benefits, in the confidence that the risk is one that is worth taking and that it is being properly controlled.

The "reasonably practicable" test only addresses the question of whether a risk is being adequately controlled according to the principles of "equity" and "utility". It cannot answer the question of whether a risk is worth taking by society.

The primary duty of a PCBU requires that, for instance, the employer of an astronaut travelling on the world's first manned expedition to Mars has the same primary duty as the employer of the person who will mow the astronaut's lawn for the months or years that he is away in space.

Most would accept that an expedition to Mars is a much riskier activity for those involved than mowing a lawn[62], but it is also an activity that may deliver substantially greater benefits to society.

Though, what employer would send an employee off to perform a task knowing that the chance of the employee being

---

[62] It also goes without saying that any astronaut who embarks on a lengthy expedition to Mars should take appropriate measures to ensure that no one "cuts his grass" while he is away.

killed doing it is perhaps 50%?[63] The requirements of the primary duty of a PCBU and the hierarchy of risk controls might give weight to an argument that a Mars expedition not be undertaken, or be undertaken without the involvement of astronauts.

Herein lies a fundamental ethical question.

In order to progress technologically and economically, society must innovate. In doing so it must take risks (see Figure 6.2). where the health and safety of certain individuals within society may be placed at the mercy of "known unknowns" and "unknown unknowns". Progress necessitates an appetite for risk.

[63] The author was once informed by a senior risk manager from NASA that the chance of mission failure (ie. all of the astronauts die) during the first manned expedition to Mars might be 50%.

Figure 6.2: Societal Risk Appetite

"We choose to go to the moon in this decade and do the other things, not because they are easy, but because they are hard, because that goal will serve to organize and measure the best of our energies and skills, because that challenge is one that we are willing to accept, one we are unwilling to postpone, and one which we intend to win." (President John. F. Kennedy 12th September 1962).

Crew of Apollo 1 – Killed during a launch test on 27th January 1967
Edward White, Virgil "Gus" Grissom & Roger Chaffee

Crew of Apollo 11 – First landing on the moon on 20th July 1969
Neil Armstrong, Michael Collins & Edwin "Buzz" Aldrin

Society likes to celebrate itself and its heroes when its risk-taking in the pursuit of opportunity is rewarded by success. Even if society and the individuals involved agree to accept the risks associated with innovation, WH&S law may take a different view if the risk-taking is not rewarded, or results in failure, and those individuals are harmed in the process.

The point of the matter is that WH&S law poses regulatory risk to PCBUs and their officers.

Could this have the unwanted effect of inhibiting our society's appetite for pursuing opportunity and the risk-taking that has been essential to innovation and technological and economic progress?

Or could it lead to gaming of the system? For example, if an organisation cannot afford a 'reasonably practicable' risk control, would it be possible that they continue to work without it, perceiving that the odds are in their favour – that it is more likely that the adverse event will not eventuate and they will not get caught out. Perhaps, in that place where opportunity, innovation and safety coincide, it is easier for PCBUs and their officers to seek forgiveness than to gain approval.

All regulation has costs and benefits. The potential benefits of the Model Act are well documented, but less has been said about its potential costs, particularly to innovation and the technological and economic progress that it brings. Perhaps it is a case that the negative impacts of the Model Act are themselves "unknown unknowns".

References

The following references were used in the writing of this book. Readers may find them useful in providing elaboration of the concepts within.

Textbooks

Johnstone, R. and Tooma, M., 2012, *"Work Health and Safety Regulation in Australia – The Model Act"*, The Federation Press.

Australian Acts and Regulations

Corporations Act 2001 (Cth).

Model Work Health and Safety Act 2011 (Cth).

Model Work Health and Safety Regulations 2011 (Cth).

Work Health and Safety Act 2011 (ACT).

Work Health and Safety Regulations 2011 (ACT).

Work Health and Safety Act 2011 (QLD).

Work Health and Safety Regulations 2011 (QLD).

Work Health and Safety Act 2011 (NSW).

Work Health and Safety Regulations 2011 (NSW).

Work Health and Safety (National Uniform Legislation) Act 2011 (NT).

Work Health and Safety (National Uniform Legislation) Regulations 2011 (NT).

Work Health and Safety Act 2012 (SA).

Work Health and Safety Regulations 2012 (SA).

Work Health and Safety Act 2011 (Tas).

Work Health and Safety Regulations 2011 (Tas).

Work Health and Safety Bill 2014 (WA) (draft legislation)

Health and Safety Reform Bill 2014 (New Zealand) (draft legislation)

Regulatory Guidelines

Comcare, *2012, "Guidance for Officers in Exercising Due Diligence"*.

New South Wales Department of Planning, 2008, *"Risk Criteria for Land Use Planning"*.

Safe Work Australia, 2012, *"Guide to the Model Work Health and Safety Act"*.

Safe Work Australia, 2011, *"Interpretative Guideline "The Meaning of Reasonably Practicable"*.

Safe Work Australia, 2011, *"Interpretative Guideline "The Health and Safety Duty of an Officer Under Section 27"*.

Safe Work Australia, 2011, *"Interpretative Guideline "The Meaning of Person Conducting a Business or Undertaking"*.

United Kingdom Health and Safety Executive, 2001, *"Reducing Risk, Protecting People - HSE's decision-making process"*, HSE Books.

Workcover NSW, 2011, *Fact Sheet "PCBUs, Workers and Officers"*.

Legal Cases

Australian Iron and Steel Pty Ltd v Environment Protection Authority [1992] 29 NSWLR 497.

Bolton v Stone [1951] AC 850.

Donoghue v Stevenson [1932] UKHL 100.

Edwards v National Coal Board [1949] All ER 743 CA.

Kirk v Industrial Relations Commission [2010] HCA 1.

Slivak v Lurgi [2001] HCA 6.

Turner v State of South Australia [1982] 42 ALR 669.

Wyong Shire Council v Shirt [1980] 146 CLR 40.

Australian and International Standards

AS/NZS ISO31000:2009 Risk Management – Principles and Guidelines.

AS/NZS4801:2001 Occupational Health and Safety Management Systems – Specification with Guidance for Use.

AS/NZS4804:2001 Occupational Health and Safety Management Systems –General Guidelines on Principles, Systems and Supporting Techniques.

AS(IEC)61508 Functional Safety of Electrical / Electronic / Programmable Electronic Safety-related Systems.

CENELEC EN50126 Specification and Demonstration of Railway Reliability, Availability, Maintainability and Safety (RAMS).

Energy Networks Association of Australia (ENA) Guideline EG-0 Power System Earthing Guide.

IEC31010:2009 Risk Management – Risk Assessment Techniques.

Institution of Gas Engineers and Managers, 2010, IGEM SR/15 - Integrity of Safety-related Systems in the Gas Industry, Edition 5.

NSW Department of Primary Industries, 1997, MDG1010 - Risk Management Handbook for the Mining Industry.

NSW Department of Trade and Investment, 2011, MDG1010 - Minerals Industry Safety and Health Risk Management Guideline.

SA/SNZ HB436:2013 Risk Management Guidelines – Companion to AS/NZS ISO31000.

SA/SNZ HB89:2013 Risk Management – Guidelines on Risk Assessment Techniques

Articles, Reports and Papers

Australian Coal Association Research Programme (ACARP) Research Paper C15011, 2009, *"Acceptable Risk: A Matter of Establishing Adequate Control"*.

Australian Coal Association Research Programme (ACARP) Research Paper C23007, 2015, *"Selection and Optimisation of Risk Controls"*.

McNicol, James., 2001, *"Have Australia's Major Hazard Facilities Learned from the Longford Disaster?"*, IEAust Public Policy Unit.

Minerals Industry Safety and Health Centre (MISHC), 2005, *"Final Report to Queensland Resources Council (QRC) on underlying Causes of Fatalities and Significant Injuries in the Australian Mining Industry"*.

Safe Work Australia, 2014, *"Work-related Traumatic Injury Fatalities, Australia 2013"*.

Websites

http://www.hse.gov.uk

http://www.legal-dictionary.thefreedictionary.com

http://www.lexisnexis.com

http://www.safeworkaustralia.gov.au

http://www.saiglobal.com

http://www.standards.org.au

http://www.sixthformlaw.info

http://www.unistudyguides.com

Index